AGRICULTURAL SYSTEMS IN TAMIL NADU

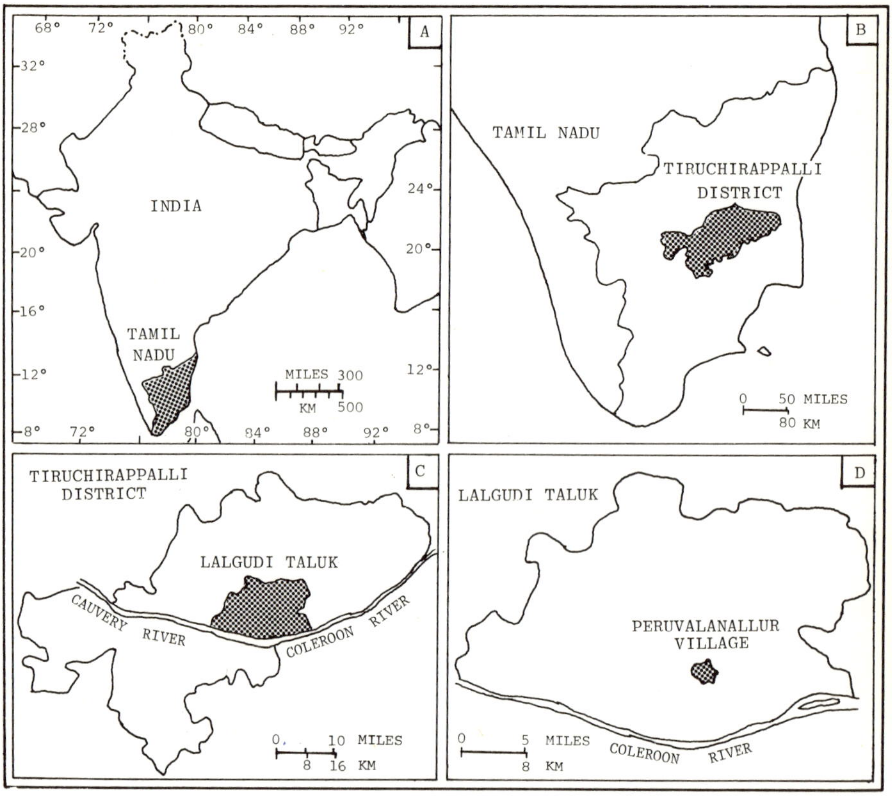

Location of Tamil Nadu State in India (A), Tiruchirappalli District in Tamil Nadu (B), Lalgudi Taluk in Tiruchirapalli District (C), and Peruvalanallur Village in Lalgudi Taluk (D). For names and locations of revenue villages in Lalgudi Taluk see table 1 and figure 4.

AGRICULTURAL SYSTEMS IN TAMIL NADU

A Case Study of Peruvalanallur Village

by
Yoshimi Komoguchi
Komazawa University

THE UNIVERSITY OF CHICAGO
DEPARTMENT OF GEOGRAPHY
RESEARCH PAPER NO. 219

1986

Published 1986 by The Department of Geography
The University of Chicago, Chicago, Illinois

Library of Congress Cataloging-in-Publication Data

Komoguchi, Yoshimi, 1937-
Agricultural systems in Tamil Nadu.

(Research paper / the University of Chicago, Department of Geography ; no. 219)
Bibliography: p. 173
1. Land tenure--India--Peruvalanallur. 2. Land use, Rural--India--Peruvalanallur. 3. Agriculture--Economic aspects--India--Peruvalanallur.
4. Peruvalanallur (India)--Rural conditions.
I. Title. II. Series: Research paper (University of Chicago. Dept. of Geography) ; no. 219.
H31.C514 no. 219 [HD880.P47] 910 s 86-6978
ISBN 0-89065-123-X [338.1'0954'82]

Research Papers are available from:

The University of Chicago
The Department of Geography
5828 S. University Avenue
Chicago, Illinois 60637-1583
Price: $10.00; $7.50 series subscription

To My Teachers:

Professor Jiro Yonekura
(A Cultural Geographer)

and

Professor Kasuke Nishimura
(A Physical Geographer)

CONTENTS

LIST OF ILLUSTRATIONS

LIST OF TABLES

ACKNOWLEDGMENTS

This monograph is a partial fulfillment of my study in the Department of Geography at the University of Chicago. My warmest thanks go to the professors and friends who have seen me through my research at the University.

In relation to my present study of a South Indian village, I am especially obliged to two professors: Professor Norton Ginsburg, who guided me with endless patience in every stage of my South Indian Studies through its many grandeurs and complexities, and Professor Chauncy Harris, who encouraged and helped my study with many ideas and suggestions.

In the Indian phase of my research, no one was more helpful and encouraging than Professor Tadahiko Hara of the Institute for Studies of Languages and Cultures of Asia and Africa (ISLCCA) in Tokyo, who organized the South Indian Research Project under which I was one of the research members. I owe gratitude to Professor A. Ramesh and Dr. S. Subbiah of the Department of Geography at the University of Madras for their cooperation with my study during my stay in South India. Special thanks go to the various Indian governments and their officials, without whose help this research could not have been accomplished.

My thirteen months' field research in South India was conducted in two phases: first from September 1979 to March 1980, second from October 1981 to March 1982. My sincere thanks go to my three research assistants who worked tirelessly and professionally for me not only during my field research period in the two phases but also for the period between the two phases: Messrs. R. Chandrasekaran, R. Rajendran, and K. Annamalai.

While living in the intensively studied village, Peruvalanallur of Tiruchirappalli Dirstrict, during both periods of the study, I got incredible cooperation from the villagers who responded to my repeated interviews in a friendly way. My warmest thanks go to each of such villagers. In Peruvalanallur, I am especially obliged to Mr. Raja Chidambaram, who not only guided me through various aspects of the village life and history but also provided me with some important historical materials including the field maps of the village published in 1893.

In the final stage of this research, I was much obliged to the following persons: Miss Hisako Koshiishi and Mr. Toyokazu Kawamoto for their data arrangement and drafting maps, Miss Yoko Ogino (Komazawa University) for her typing and secretarial work, and Mr. Charles Cushman (Tokyo University) for proof-reading of my English.

To achieve this study, I have received financial assistance from several different sources: in the USA, a Fulbright Fellowship, and Advanced Research Fellowship of the East-West Center, a research grant from the Asia Foundation, and a University of Chicago Fellowship; and in Japan, the Overseas Research Grant from the Japanese Ministry of Education and a research grant from Komazawa University. I would like to express my sincere gratitude to the above institutes and governmental agencies.

My deepest gratitude and affection goes to my family members: my mother-in-law, Mrs. Shige Doi; my wife, Chieko; and my daughters, Makiko and Minako, who have supported in one way or another.

Last but not least, I owe gratitude to my two great teachers, Professor Jiro Yonekura (a cultural geographer) and Professor Kasuke Nishimura (a physical geographer), who have warmly seen me through my study activities for many years. Thus, I would like to dedicate this essay to my teachers.

Yoshimi Komoguchi

Tokyo, April 1985

CHAPTER I
INTRODUCTION

Peruvalanallur village is located in southern India in the state of Tamil Nadu, in Tiruchirappalli District, in Lalgudi Taluk (see Frontispiece). The village consists mostly of irrigated wet lands. Agriculture is favored by a year-around growing season, by a double rainy period, one during the southwest monsoon of June-September and the other during the northeast monsoon of October to December, and by extensive irrigation systems based on waters from the Cauvery River and particularly one of its distributaries, the Coleroon River, stored in two large tanks in the village, and by local water both from deep tubewells in the drier and higher northern part of the village and from shallow tubewells in the lower and wetter southern part of the village. In spite of the double monsoon the average annual amount of precipitation is only about 700 mm or 28 inches, a modest amount for such a warm area, but enough to make possible some crops on the dry unirrigated lands. The main agriculture of the village, however, is based on the irrigated lands, which in recent years have become predominantly double cropped. The double cropping had been made possible by an extension of the irrigation season through more extensive irrigation works.

Framework of the Study

The alluvial plains of the Cauvery River basin in Tamil Nadu are noted as one of the more prominent agricultural regions in India. The development of agriculture in this region has been largely dependent upon irrigation systems throughout its history.[1] However, on some terraces and slopes surrounding the Cauvery plains, there are extensive areas of upland where water for agriculture is derived directly from monsoon rains.

---

1. Noboru Karashima, "Land Revenue Assessment in Cola Times as Seen in the Inscriptions of the Thanjavur and Gangaikondacholapuram Temples," *Studies in Socio-Cultural Change in Rural Villages in Tiruchirappalli District, Tamil Nadu, India,* no. 1 (Tokyo: Institute for the Study of Languages and Cultures of Asia and Africa [ISLCAA], November 1980), pp. 35-50.

This study is fundamentally concerned with the rural community and agricultural systems of South India, with specific reference to the variations, both geographical and functional, found in their socio-economic characteristics. It deals primarily with a selected rural community, Peruvalanallur village (fig. 1).

Many people have often discussed rural communities in relation to the development of less-developed countries, but paradoxically a basic understanding of their structural patterns and processes has yet to be attained. This failure may be due to the fact that only a relatively few comprehensive studies at the village level are presently available. The Indian situation in this respect does not seem to be an exception.[2] Moreover, among the available village-level studies in India, the geographers' contributions have been relatively fewer than they ought to be, even though the analytical method based on conventional ecological and spatial concepts can be applied effectively to the study of some important aspects of the rural community and its development. This study exemplifies this line of analysis.

However, the study also has its own specific objectives: first, to clarify the structural and spatial patterns of the rural community with special reference to its agricultural activities and to measure the degree of association and integration among villagers whose socio-economic backgrounds seem to vary to a great extent; second, to investigate the recent changes in the socio-economic aspects of the community; and by extension, third, to identify some important elements responsible for its modernization. In order to meet these objectives, the research has focused on the following four topics: (1) land-uses and their associations, (2) landownership, (3) occupational specialization and labor organization, and (4) land tenure. A chapter is devoted to each topic.

It is well known that there are several caste *(jati)* groups in any given rural community in India and that they form a socio-economic stratification. There are (1) a few dominating groups, (2) several dependent castes, and (3) between the above two poles, a series of caste groups which have some range of socio-economic variations within them. Thus, besides the individual and household, each caste group is regarded as an analytical unit in this study. Another important analytical unit adopted here is the

---

2. Shanmugam P. Subbiah, "Rural Base in a South Indian Village: A Study into Its Structural and Spatial Patterns in Mahizambadi Village of Tamil Nadu," *Studies in Socio-Cultural Change in Rural Villages in Tiruchirappalli District, Tamil Nadu, India,* no. 4 (Tokyo: ISLCAA, August 1981), p. 1.

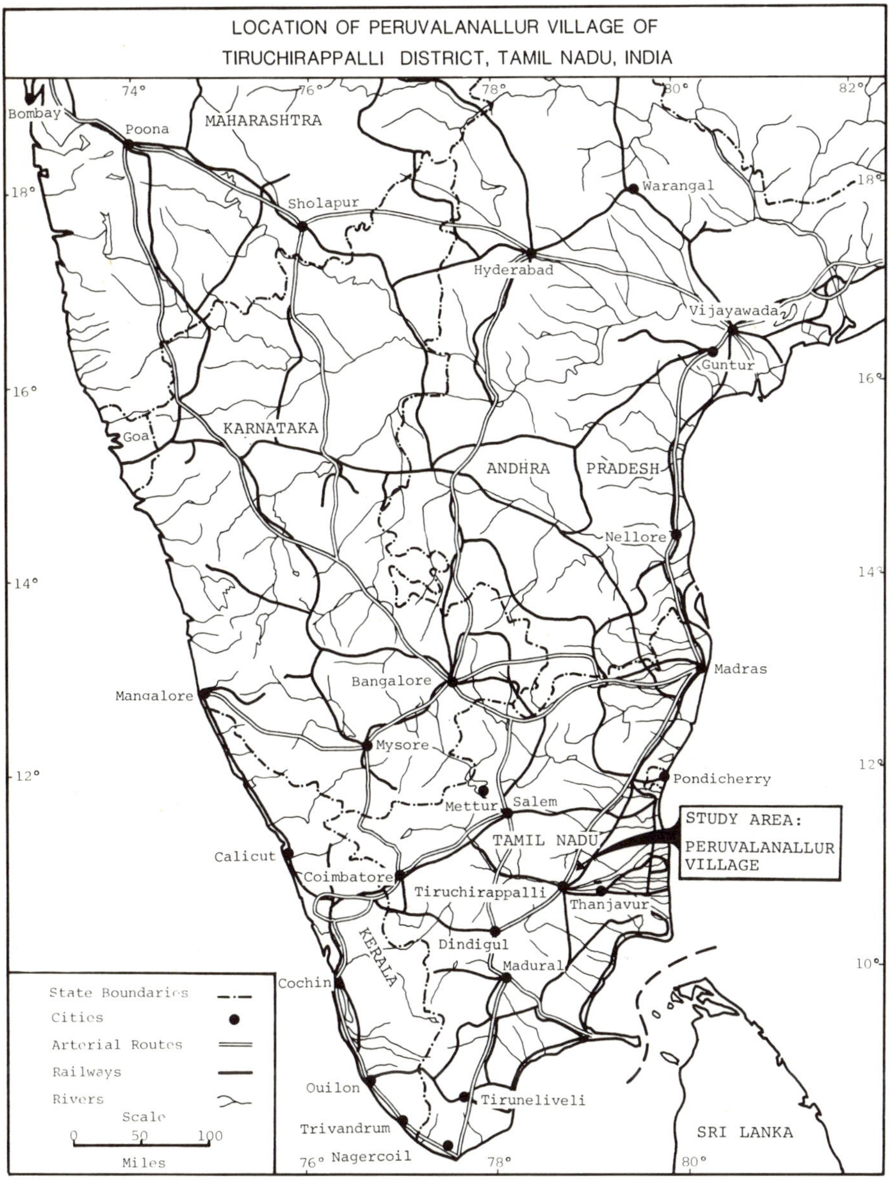
LOCATION OF PERUVALANALLUR VILLAGE OF
TIRUCHIRAPPALLI DISTRICT, TAMIL NADU, INDIA
74°
76°
78°
80°
82°
18°
16°
14°
12°
10°
Bombay
Poona
MAHARASHTRA
Sholapur
Warangal
Hyderabad
Vijayawada
Guntur
KARNATAKA
Goa
ANDHRA
PRADESH
Nellore
Madras
Bangalore
Mangalore
Mysore
Pondicherry
Mettur
Salem
TAMIL NADU
STUDY AREA:
PERUVALANALLUR
VILLAGE
Calicut
Coimbatore
Tiruchirappalli
Thanjavur
KERALA
Dindigul
Madurai
Cochin
Quilon
Tirunelveli
Trivandrum
Nagercoil
SRI LANKA
State Boundaries
Cities
Arterial Routes
Railways
Rivers
Scale
0
50
100
Miles

Fig. 1

class of individuals and households based on the different sizes of their landholdings and operational farms, because land is still regarded as the most important factor of agricultural production. Therefore, our description and analysis of socio-economic aspects in this study will often be based on the caste, household, and individual levels under a specifically classified size of landholdings.

Some of the characteristics of the problems involved in this study and the procedures to analyze them are discussed here. The chapter on land-uses and their associations mainly discusses seasonal cropping patterns and their conditions within an ecological framework. It takes into account such physical elements as climate, field configurations, and the relative location of the fields with respect to the water sources on the one hand and technological elements such as the patterns of crop-breeding and selection, manuring, and water supply and control on the other.

Water supply and control is one of the key factors for any type of agriculture in a given region. Specifically, the availability of a water supply for irrigation is an especially important factor for the studied village and its surrounding area whose agricultural activities involve the cultivation of different types of paddy,[3] sugar cane, bananas, several kinds of pulses and millets, and other field crops under the tropical savanna type of climatic conditions. This general proposition may appear to be self-evident, but it also conceals some complexities and nuances.[4] Therefore, it has been necessary to determine the areal extent to which crop lands under different physiographic conditions retain a sufficient or desirable volume of water and to discern how effectively different types of paddy and other crops are cultivated in accordance with the available water supply.

Agricultural lands in Tamil Nadu have been officially classified into two major categories, wet (lower) and dry (higher) lands, of which the former are subdivided into "single" and "double" cropping lands. However, the terms "single" and "double" should not be taken literally. This distinction has been made based on whether one or two of the major crops, such as different types of wet-paddy and sugar cane,[5] can possibly be cultivated on a particular piece of

---

3. The term "type of paddy" as used here does not mean a variety of rice plant. A "type" is a number of varieties of paddy classified together by their common cropping season.

4. Clifford Geertz, *Agricultural Involution: The Process of Ecological Change in Indonesia* (Berkeley: University of California Press, 1966), p.31. See also D.H. Grist, *Rice* (London: Longmans Green, 1959), pp. 28-29.

5. The land used for sugar cane is officially regarded as a "double"

land. In fact, besides the major crops, some of the gram, gingelly (sesames), and other field crops have been extensively cultivated in the "single" and "double" cropping lands in the same crop year in and around the studied village.

What is important about these categories of land in relation to this study is that the change of areas from dry to "single-wet" and from "single-wet" to "double-wet" lands in a particular area during a given period may reveal, to some extent, the development of the irrigation system and, by extension, of agriculture itself. Thus, in this connection, the following questions are raised: (1) to what areal extent have the individual categories of land been respectively changed during the specific periods in the studied village, (2) where have such changes occurred, and (3) what were/are the important sources of agricultural water? To answer these questions the available historical records and more recent data obtained during the author's field work are carefully examined on a field-to-field basis, and they are compared with each other.

Along with the inquiries into the historical background of the crop lands, an extensive investigation has been placed upon the present land-uses and their crop associations in the studied village. The development of the irrigation systems seems to have been conducive to reducing the emphasis on "traditional" practices of agriculture in general. More specifically, the development may not only allow a wider range of crop selection in different seasons, but also stimulate more farmer incentives toward the adoption of modern agricultural inputs and the high-yielding varieties. In this respect, the study area, Peruvalanallur, which admittedly is one of the most advanced villages in agriculture among the revenue villages in Lalgudi Taluk,[6] will be examined.

Various patterns of assigning different crops to specific plots of land, to their rotation, and to the cropping seasons must not be thought of as static or unchangeable. What differentiates regions, villages, and even the sections of a village is the relative emphasis placed upon a particular type, or a particular combination of types, of crops. In addition, it must be stressed that the gross production and net production per unit of land for particular types of crops vary from place to place. Thus, these variations have been compared and accounted for by benefit-cost analyses of the individual crops. The analyses help clarify those

---

cropping area, because the same land can be used for the cultivation of two different types of wet-paddy in a year.

6. The village is so recognized not only by the officials, but also by its own residents and the neighboring villagers.

factors which bring about specialized land utilization and combinations of crops. It goes without saying, however, the the benefit-cost analysis cannot explain all the decisions actually made by the respective cultivators, and that analysis itself can not go beyond the range of alternatives which are immediately evident to the cultivators and directly observed by the author.

Aspects of landownership are mainly discussed as follows: (1) the distribution of lands among the households of the studied village as ownership and operational farms, (2) the spatial distribution of lands under the villagers' ownership, and (3) the transfer of land. In most villages in South India the available agricultural land in a village is partly owned by outsiders, but mostly by village residents. Therefore, the investigation of landownership has been pursued not only in the intensively studied village, Peruvalanallur, but also in its related villages.

It is often emphasized that the available lands of the villages in South India are concentrated to a great extent in a limited number of households which belong to a particular caste group. Thus, this aspect has been examined in the studied village in a manner that shows the variation in sizes of landholdings among its households by caste. The expected variation seems to be one of the major elements in describing the various patterns of socio-economic activities, especially those of land tenure, in the rural community. Because of the results of land transactions under certain types of land tenure, the size of each household's landholdings and the size of the operational farm are not necessarily the same. Thus, the distribution of operational farms among the households in the village as a whole must be clarified. Then, its distribution pattern is compared with that of landownership. The comparison may help to characterize classes or groups of households, defined by the size of landholding, in relation to the actual land they individually operate, and thus perceive trends in land tenure in the different class groups, i.e., the complex attitudes among groups toward owning land as opposed to managing it.

Concerning the basic data employed, the landholdings of individual households have been obtained by carefully examining various sources including the official records, the villagers' personal documents, and the author's own materials gathered from intensive interviews. In fact the available plots of land in the studied village have been individually examined as to their owner's names, areal extent, distinction between dry, "single-wet," and "double-wet" lands, and their location. The same kind of survey has been also conducted in neighboring villages.

As indicated above, the residents of the studied village hold lands not only in their own village but also in other villages. Conversely, other village residents own a sizeable extent of land in the studied village. What is of concern here, then, is the spatial distribution of landownership. In this connection, it is necessary to identify and/or determine the following: (1) the location of other individual villages and the corresponding areas of lands owned by the studied village residents, (2) the location of individual fields and their corresponding owners in the studied village, (3) the location and areas of lands in the studied village owned by outsiders in different villages. These materials help clarify some important aspects such as the relationship between the spatial distribution of land and farming activities, the distribution of landownership among the different caste groups, and the areal extent or limits of the inter-village relationships of landownership. The inter-village distribution of landownership will be related to farm management, property inheritance and division, and different types of migration.

After the examination of land-uses and their associations, and of landownership and its implications, the study turns to an analysis of occupational specialization in the village in relation to male and female labor, economic class, and caste groups, and to labor organization.

The main part of the study then concludes with an extensive discussion of the various types of land tenure found in the village, particularly the long-term continuing fixed rent tenure system called *kuttagai* and the short-term usufructuary mortgage tenure called *otti*. These are placed within the general setting of land tenure and tenant regulations in Tamil Nadu state

Sources of Data

The field work for this study was conducted in two phases: first, during September 1979-March 1980; second, during October 1981-March 1982. The basic family census of all of the 874 households available in the studied village was taken by the face-to-face interview method using prepared questionaires. Of course, the interview method alone has limitations for collecting reliable information or data on some delicate aspects such as property holdings and related matters, specifically on landownership and land tenure. However, thanks to the highly developed Indian bureaucratic system, we had access to various kinds of official documents at the revenue village level. The following materials were important sources for study: (1) the *First Settlement Register* (1864), (2) the *First Re-settlement Register* (1898), (3) the *Second Re-settlement Register* (1927) with the current documents attached, (4) the *Field Map Books* (maps for each field with its sub-division corresponding to the *Second Re-settlement Register* with its current changes), (5) the *Ten-One Chitta* (the record of landownership under their individual names and identification numbers attributed to the original landowners), (6) the *Adangals* (the annual records for land-uses in each field and its sub-divisions), and (7) the *Tenant Registers* (1972).

In order to make the data current for this study, most of them were up-dated to the end of February 1980. In the process of the adjustment, the important items for this study were carefully checked and drawn on maps whenever possible, although only part of them are described in this paper.

CHAPTER II
THE STUDY AREA

Peruvalanallur Village of Lalgudi Taluk is located in the Tiruchirappalli District of Tamil Nadu. The major part of Tamil Nadu is climatologically characterized as savanna-type with two different monsoonal rains, both the southwest monsoon of June-September, which brings rain to most of India, and the northeast monsoon of October to December, which brings rain to the Coromandel coast of South India, mainly in Tamil Nadu. The annual (and monthly) rainfall varies greatly, however, and shows a cyclical feature every four or five years. Figure 2 shows the average annual rainfall for the 50 years (1921-1971) in the Tiruchirappalli (also called Tiruchy) District. The average annual rainfall in the district amounts to 806 mm, of which about 49 percent occurs during the northeast monsoon period (October-December), about 31 percent during the southwest monsoon period (June-September), and about 20 percent during the months of January-May. The figure reveals that the rainfall is the heaviest in the east and that it gradually decreases towards the west. Thus the average annual rainfall near Jayankodam in the east is 1,086 mm, which is about 35 percent above the average for the district, while the average rainfall at Palaviduthi, located about 50 km west-southwest of Tiruchy, is only 477 mm, which is about 41 percent below the average annual rainfall for the district.

Figure 3 shows the pattern of drainage and elevation of Lalgudi Taluk of which the study area is a part. Within the *taluk,* the higher land (350-400 feet above sea level) appears in its northern and northwestern sections from which the land slopes in a southerly and southeasterly direction to the Coleroon River (a distributary of the Cauvery River). In terms of agricultural activities, the *taluk* can be divided into two physiographic zones: the wet zone and the dry zone. The wet zone corresponds to the alluvial plains distributed narrowly along the Coleroon River. It should be noted that the agricultural activities in this zone have

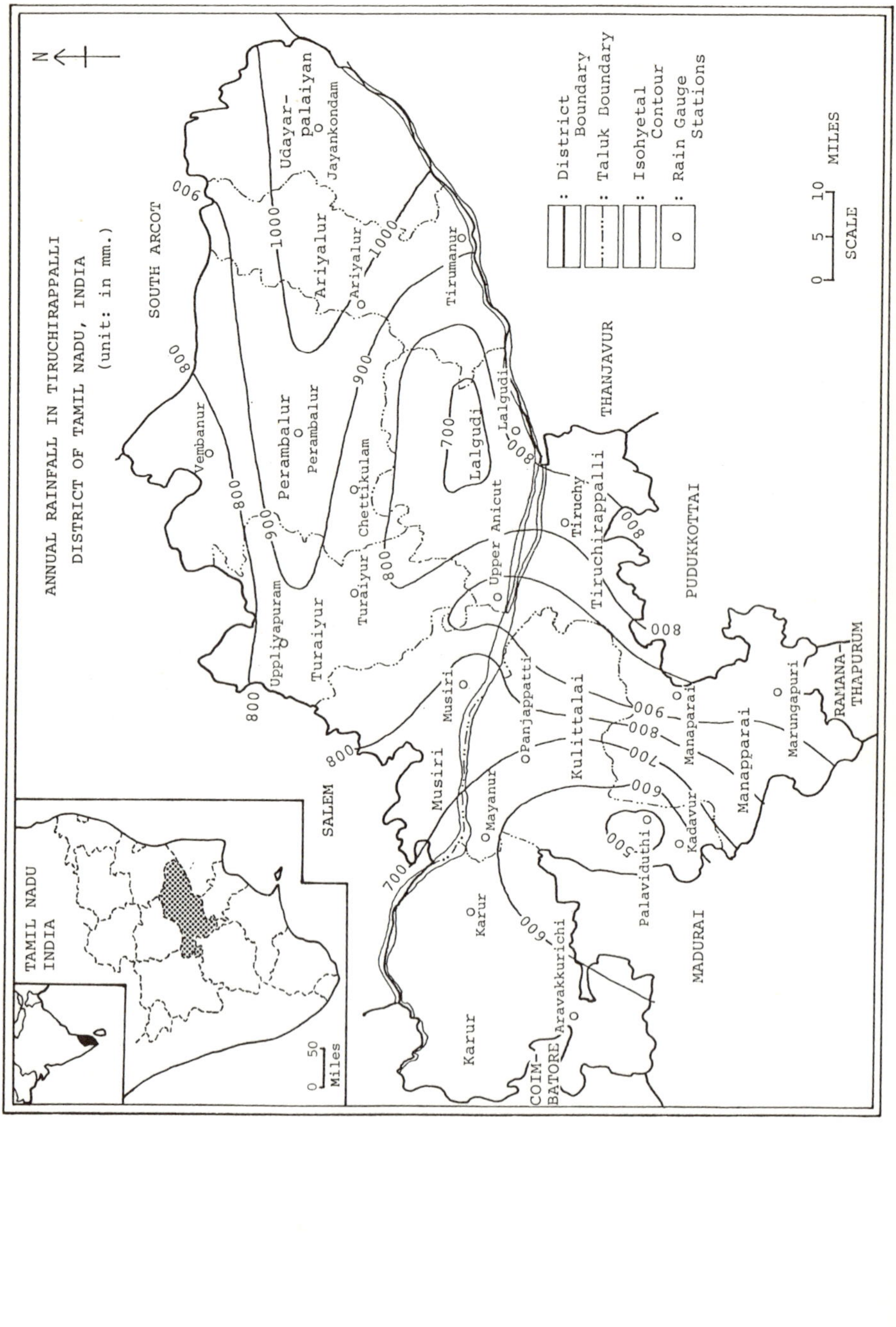

ANNUAL RAINFALL IN TIRUCHIRAPPALLI
DISTRICT OF TAMIL NADU, INDIA
(unit: in mm.)
N
TAMIL NADU
INDIA
0 50
Miles
SOUTH ARCOT
THANJAVUR
PUDUKKOTTAI
SALEM
MADURAI
COIM-
BATORE
RAMANA-
THAPURUM
Udayar-
palaiyan
Jayankondam
Ariyalur
Ariyalur
Tirumanur
Perambalur
Perambalur
Vembanur
Chettikulam
Lalgudi
Lalgudi
Upper Anicut
Tiruchy
Tiruchirappalli
Turaiyur
Turaiyur
Uppliyapuram
Musiri
Musiri
Panjappatti
Kulittalai
Manaparai
Manapparai
Marungapuri
Mayanur
Kadavur
Palaviduthi
Karur
Karur
Aravakkurichi
1000
900
800
700
600
500
: District Boundary
: Taluk Boundary
: Isohyetal Contour
: Rain Gauge Stations
0 5 10 MILES
SCALE

Fig. 2

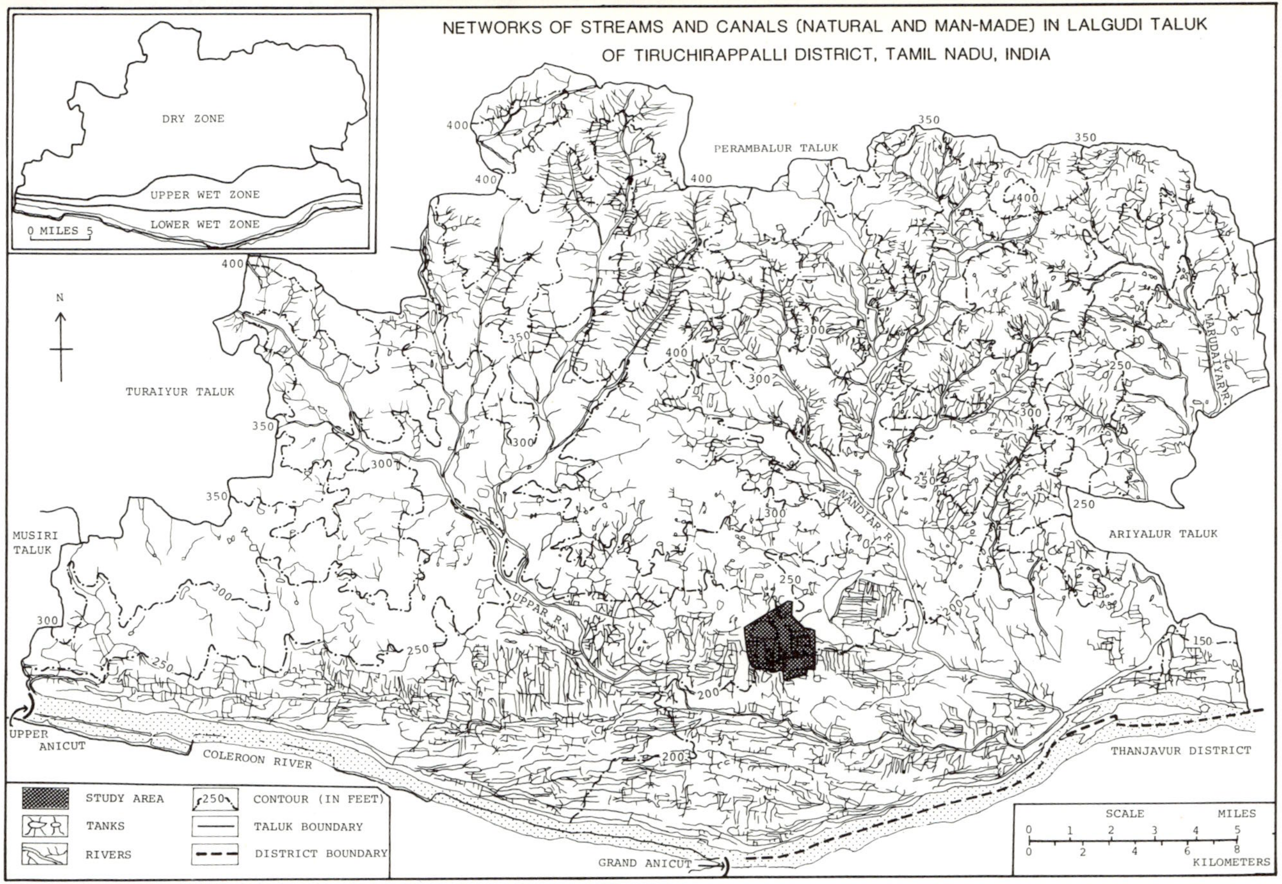
NETWORKS OF STREAMS AND CANALS (NATURAL AND MAN-MADE) IN LALGUDI TALUK
OF TIRUCHIRAPPALLI DISTRICT, TAMIL NADU, INDIA
DRY ZONE
UPPER WET ZONE
LOWER WET ZONE
0 MILES 5
N
PERAMBALUR TALUK
TURAIYUR TALUK
MUSIRI TALUK
ARIYALUR TALUK
THANJAVUR DISTRICT
MARUDAIYAR R.
NANDIAR R.
UPPAR R.
COLEROON RIVER
UPPER ANICUT
GRAND ANICUT
STUDY AREA
TANKS
RIVERS
CONTOUR (IN FEET)
TALUK BOUNDARY
DISTRICT BOUNDARY
SCALE
MILES
KILOMETERS

Fig. 3

been largely dependent upon the development of the irrigation systems. There are several irrigation channels in the wet zone, which are connected with the Upper Anicut of the Cauvery River. The wet zone can be further divided into two physiographic sub-zones, that is, a "lower wet" zone and an "upper wet" zone. The availability of irrigation water is more stable in the former than in the latter. The study area belongs to the "upper wet" zone.

The vast area of the dry zone extends to the north of the upper wet zone and is highly dissected along the non-perennial jungle streams (*varis*). The water for agricultural use in the dry zone is mostly dependent upon the direct monsoon rains during October-December. However, in many villages of the dry zone in the *taluk,* there are some "wet land" areas being irrigated from different sources: *eris* (tanks), traditional wells, and ground water which the irrigation techniques themselves have been changing recently from traditional *aetram-eravai* (slope-lifting by people) and *kavalai* (skin-bag-lifting by a pair of bullocks) to ones using inanimate energy.

Peruvalanallur is one of the 126 revenue villages under the Lalgudi Taluk, Tiruchy District (number 63 in table 1 and fig. 4).[1] The village consists of two different settlements: (1) Peruvalanallur as the major one and (2) Pallapuram as the minor. These two settlements are organized under an integrated administrative unit which is mainly responsible for various types of revenue collections, the administration of which is undertaken by an appointed village *munshif* (head) and a few of his assistants who are all under the higher control of the Lalgudi Taluk.[2] Each of the above settlements, however, has its own *panchayat,* the smallest administrative unit for various aspects of welfare and general village development and control.[3] An intensive study has been made here only of the major settlement, the Peruvalanallur part of the village.

The study area, lying on the left bank of the Coleroon River, is located some 26 km northeast of Tiruchy, the district capital of the same name, and some 6 km east-northeast of Lalgudi, a "village

---

1. The village numbers shown in table 1 and figure 5 are not arbitrary, but are the fixed official reference numbers assigned to the respective villages, which first appeared in the *First Re-settlement Register* published in 1898.

2. However, in 1980 this system was abolished, and most of the administration under the previous system merged with the *panchayat* administration.

3. There are many such revenue villages which have two or more *panchayats* within each of them. In these cases, the names of the revenue villages are taken from their respective major settlements, and of course the major settlements themselves have their own *panchayats* as in the case of Peruvalanallur.

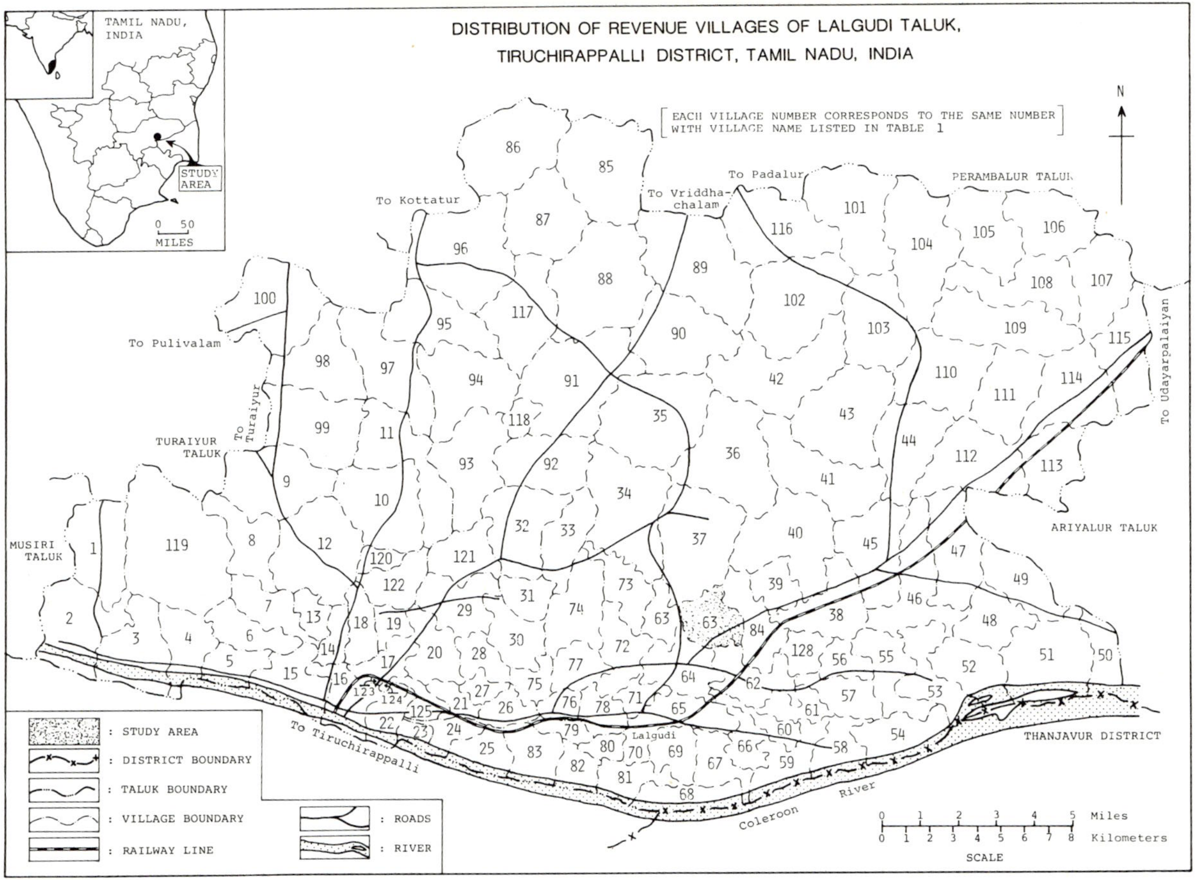
DISTRIBUTION OF REVENUE VILLAGES OF LALGUDI TALUK,
TIRUCHIRAPPALLI DISTRICT, TAMIL NADU, INDIA
EACH VILLAGE NUMBER CORRESPONDS TO THE SAME NUMBER WITH VILLAGE NAME LISTED IN TABLE 1
N
TAMIL NADU, INDIA
STUDY AREA
0 50 MILES
To Kottatur
To Vriddha-chalam
To Padalur
PERAMBALUR TALUK
To Pulivalam
To Turaiyur
TURAIYUR TALUK
MUSIRI TALUK
To Udayarpalaiyan
ARIYALUR TALUK
THANJAVUR DISTRICT
Lalgudi
To Tiruchirappalli
Coleroon River
: STUDY AREA
: DISTRICT BOUNDARY
: TALUK BOUNDARY
: VILLAGE BOUNDARY
: RAILWAY LINE
: ROADS
: RIVER
0 1 2 3 4 5 Miles
0 1 2 3 4 5 6 7 8 Kilometers
SCALE

Fig. 4

TABLE 1

REVENUE VILLAGE LIST OF LALGUDI TALUK, TIRUCHIRAPPALLI DISTRICT, TAMIL NADU, INDIA

1	Cholaganallur	33	Mahilambadi	65	Manakkal	97	Edumalai
2	Kariyamanikkam	34	Thachankuruchi	66	Adigudi	98	Sirugudi
3	Kiliyanallur	35	Reddimangudi	67	Satthamangalam	99	Siruppattur
4	Melappattu	36	Kannagudi	68	Kukur	100	Omandur
5	Tiruvasi	37	Kumulur	69	Mummudicholamangalam	101	Teranipalaiyam
6	Kovattagudi	38	Vellanur	70	Labishekapuram	102	Nambukurichi
7	Alagiyamanavalam	39	Sangendi (Nanjai)	71	Sirudaiyur	103	Sirukalapur
8	Tirampalaiyam	40	Sangendi (Punjai)	72	Thirumangalam	104	Garudamangalam
9	Tiruvallarai	41	Vandalaigudalur	73	Pudur Uthamanur	105	Saradhamangalam
10	Ayyampalaiyam	42	Peruvalappur	74	Neykuppai	106	Kannanur
11	Tattamangalam	43	Kanakkiliyanallur	75	Kilperungavur	107	Orattur
12	Punnampalaiyam	44	Tappy	76	Nerunjilakudi	108	Chathurpagam
13	Ulundangudi	45	Pullambadi	77	Nagar	109	Malvay
14	Mannachanallur	46	Venkatachalapuram	78	Angarai	110	Varakuppai
15	Madavaperumal Kovil	47	Kovandakurichi	79	Pambaramsuthi	111	Melarasur
16	Bhishandar Kovil	48	Allampakkam Ullur	80	Seshasamudram	112	Kallakudi
17	Kuthur	49	Pudur Palayam	81	Edaiyathumangalam	113	Muduvattur
18	Melsidevimangalam	50	Alambadi	82	Javandinathapuram	114	Kilarasur
19	Vengangudi	51	Virahalur	83	Tirumanamedu	115	Kallagam
20	Madakkudi	52	Nattamangudi	84	Edangimangalam	116	Uttattur
21	Pudukkudi	53	Alangudi Mahajanam	85	Sidevimangalam	117	Tiruppattur
22	Talakkudi	54	Ariyur	86	Perakambi	118	Avaravalu
23	Appadurai	55	Kalligudi (Tappay)	87	Vailaiyur	119	Tiruppangili
24	Esanakkorai	56	Sembarai	88	Shannamangalam	120	Rajampalaikam
25	T. Valavanur	57	Tinniyam	89	Neykulam	121	Samayapuram
26	Valadi	58	Kilambil	90	Periyakurukkai	122	Kalpalaiyam
27	Sirumarudur	59	Mangamapuram	91	Siruganur	123	Uthamanambi
28	V. Turaiyur	60	Jangamrajapuram	92	Konalai	124	Chinnavalakurichi
29	S. Kannanur	61	Mettuppatti	93	Ayikkudi	125	Keeramangalam
30	Marudur	62	Sirumayangudi	94	Kariyamanikkam	126	No village
31	R. Valavanur	63	Peruvalanallur	95	Palaiyur	127	No village
32	Irungalur	64	Poovalur	96	Edumalai	128	Komagudi

town,"[4] where the *taluk* head office is located (fig. 4). Being connected with the Tiruchy-Udaiyarpalaiyan Highway, the village is in a good position to reach the surrounding areas. Given a well developed bus service, it is within one hour's traveling distance to the Main Guard Gate, the commercial center of Tiruchy, and about 15 minutes from Lalgudi.

Peruvalanallur (the village) has a territory of 1,335.70 acres. The general land-use in the study area is shown in figure 5 and table 2. The residential areas of the village cluster are a little to the north of the center of the territory (figs. 5 and 6). Although the villagers are in a convenient position to communicate with neighboring villages through existing roads and pathways, the north-south (bus) road is the most important one, since it stretches from the major residential areas to the southern "gateway" and connects with the Tiruchy-Udaiyarpalaiyan Highway.

There are two big tanks for irrigation, the *mela eri* (western tank) and the *kila eri* (eastern tank), extending to the west and east of the major residential areas of the village. Being connected with the Peruvalai channel, the tanks provide water partly to the immediate surrounding fields, but mostly to their southern fields through well developed field canal networks. This kind of irrigated land is called *nanjai* (irrigated wet land), whereas the only rain-fed land on the higher elevations found mostly to the north of the Pullambadi channel is called *punjai* (non-irrigated dry land); but there are some pockets of tubewell irrigated areas on the higher lands. Of the total agricultural land (947.51 acres)[5] in the study area, the *nanjai* and *punjai* occupy 825.38 acres and 122.13 acres respectively. Peruvalanallur is usually identified as a "wet village" in this locality. As will be discussed in the next chapter, there are marked differences between the agricultural activities of the ecotypes, *nanjai* and *punjai*.

In relation to the socio-economic aspects of the study area, some important points should be noted. Table 3 shows some basic statistics for the study area by the different castes (*jati*s). The residents of Peruvalanallur consist of 33 caste groups and 874 households (3,496 persons). Of the groups, there are limited numbers of big landowners belonging to the *Reddiar*s (no. 2 in table 3), the leading caste in Peruvalanallur; they possess more than half

---

4. The local officials and people call it a "town," but it officially consists of three villages: (1) Mummudicholamangalam (no. 69), (2) Labishekapuram (no. 70), and (3) Sirudaiyur (no. 71).

5. Agricultural land here includes uncultivated areas, which are mostly used for grazing.

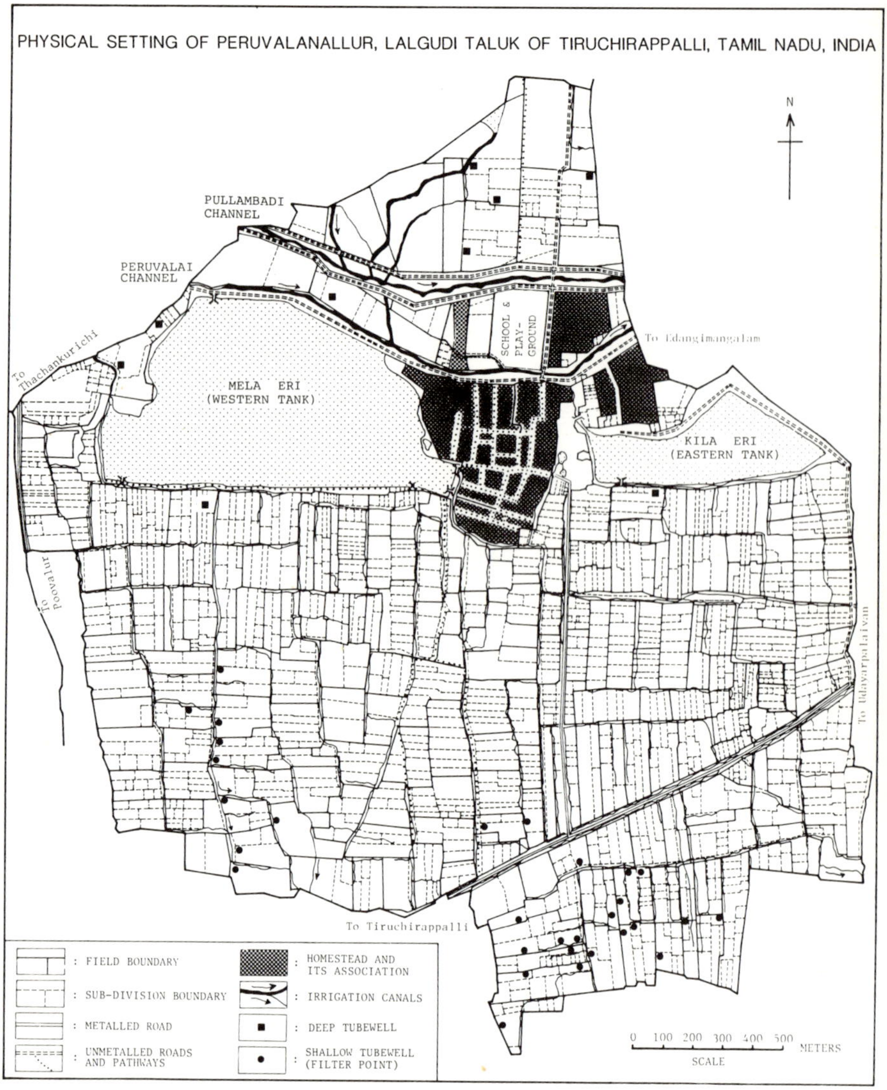
PHYSICAL SETTING OF PERUVALANALLUR, LALGUDI TALUK OF TIRUCHIRAPPALLI, TAMIL NADU, INDIA
N
PULLAMBADI CHANNEL
PERUVALAI CHANNEL
To Thachankurichi
MELA ERI (WESTERN TANK)
SCHOOL & PLAY-GROUND
To Edangimangalam
KILA ERI (EASTERN TANK)
To Poovalur
To Udavarpalaiyan
To Tiruchirappalli
: FIELD BOUNDARY
: SUB-DIVISION BOUNDARY
: METALLED ROAD
: UNMETALLED ROADS AND PATHWAYS
: HOMESTEAD AND ITS ASSOCIATION
: IRRIGATION CANALS
: DEEP TUBEWELL
: SHALLOW TUBEWELL (FILTER POINT)
0 100 200 300 400 500 METERS
SCALE

Fig. 5

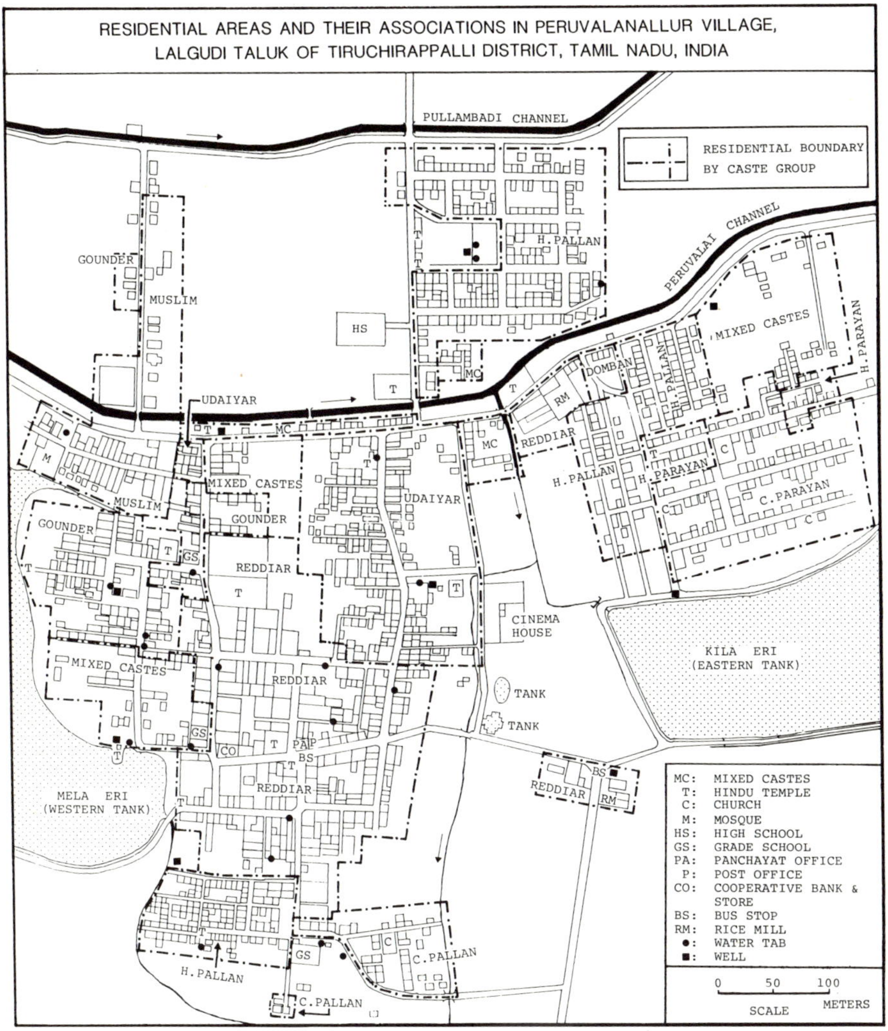
RESIDENTIAL AREAS AND THEIR ASSOCIATIONS IN PERUVALANALLUR VILLAGE, LALGUDI TALUK OF TIRUCHIRAPPALLI DISTRICT, TAMIL NADU, INDIA
PULLAMBADI CHANNEL
RESIDENTIAL BOUNDARY BY CASTE GROUP
PERUVALAI CHANNEL
GOUNDER
MUSLIM
H.PALLAN
HS
MIXED CASTES
H.PARAYAN
DOMBAN
UDAIYAR
RM
MC
M
MIXED CASTES
REDDIAR
H.PALLAN
H.PARAYAN
C.PARAYAN
GOUNDER
REDDIAR
CINEMA HOUSE
KILA ERI
(EASTERN TANK)
TANK
MIXED CASTES
MELA ERI
(WESTERN TANK)
CO
PA
P
BS
GS
C.PALLAN
MC: MIXED CASTES
T: HINDU TEMPLE
C: CHURCH
M: MOSQUE
HS: HIGH SCHOOL
GS: GRADE SCHOOL
PA: PANCHAYAT OFFICE
P: POST OFFICE
CO: COOPERATIVE BANK & STORE
BS: BUS STOP
RM: RICE MILL
●: WATER TAB
■: WELL
0 50 100
SCALE METERS

Fig. 6

TABLE 2

SUMMARY OF LAND-USE IN PERUVALANALLUR VILLAGE OF LALGUDI TALUK, TIRUCHIRAPPALLI DISTRICT, TAMIL NADU, INDIA

(1978-79)

Land-uses	Areas (in Acre)	Percentage
1. Residences and Associated Areas including Temples, Mosques, Public Buildings, etc.	98.10	7.35
2. Schools and Playgrounds	13.79	1.03
3. Agricultural Lands	947.51	70.94
(a) Wet Lands	(825.38)	(61.80)
(b) Dry Lands[1]	(122.13)	(9.14)
4. Roads and Pathways	23.13	1.73
5. *Eris* (Tanks for Irrigation) and Other Water Bodies	191.02	14.30
6. Irrigation Channels and Field Canals	61.46	4.60
7. Others	0.69	0.05
Total	1,335.70	100.00

Sources of Data:

The data consist of the *Second Re-settlement Register* (1927) with its current documents, the *Adangals* (the Annual Land-use Records) for 1978-79 and 1979-80 crop years, and the *Field Map Books* for Peruvalanallur revenue village (#63).

Although this table seems simple, the land uses and areas for each field and its sub-divisions (see Fig. 5) have been carefully checked.

Note:

1. The dry lands include the uncultivated area.

TABLE 3

SOME BASIC STATISTICS OF CASTE GROUPS IN THE PERUVALANALLUR VILLAGE, LALGUDI TALUK OF TIRUCHIRAPPALLI DISTRICT, TAMIL NADU, INDIA

(1980)

	1 No. of House-holds	2 No. of Persons	3 Owned Lands (in Acre)	4 Percent-age of (3)	5 Operation-al Lands (in Acre)	6 Percent-age of (5)
I. Forward Castes						
1. Brahman	3	23	-	-	-	-
2. Reddiar	79	325	974.38	58.82	797.90	50.07
Sub-total	82	348	974.38	58.82	797.90	50.07
II. Backward (Middle Class) Castes						
3. Udaiyar	118	411	190.92	11.52	200.59	12.59
4. Gounder	73	277	178.08	10.75	182.20	11.43
5. Muslim (Labbai)	51	241	90.44	5.46	83.50	5.24
6. Nadar	15	67	0.45	0.03	2.79	0.18
7. Achari	14	59	1.74	0.10	1.14	0.07
8. Muthuraja	14	56	6.06	0.37	19.38	1.22
9. Naidu	10	37	-	-	0.28	0.02
10. Pillai	8	26	-	-	-	-
11. Vannan (Dobi)	7	33	3.49	0.21	1.49	0.09
12. Pariyari	7	29	2.27	0.14	2.74	0.17
13. Chettiar	6	22	8.24	0.50	8.24	0.52
14. Pandaram	6	37	16.31	0.98	18.44	1.16
15. Mooppanar	4	9	1.00	0.06	0.22	0.01
16. Vellalar	3	11	-	-	-	-
17. Agampadiar	3	9	-	-	0.39	0.02
18. Muthaliar	2	11	0.10	0.01	0.10	0.01
19. Padaiyachi	2	12	-	-	-	-
20. Christian (Protestant)	2	12	1.50	0.09	-	-
21. Devar	1	4	-	-	-	-
22. Konar	1	5	-	-	-	-
23. Jangam	1	4	-	-	-	-
24. Nayar	1	5	-	-	-	-
Sub-total	349	1,377	500.60	30.22	521.50	32.73
III. Scheduled Castes (Harijans)						
25. Pallan	302	1,168	155.89	9.41	222.43	13.96
26. Parayan	27	112	4.13	0.25	6.56	0.41
27. Catholic Pallan	28	108	1.22	0.07	11.91	0.75
28. Catholic Parayan	63	286	19.81	1.19	33.24	2.08
29. Harijan Pariyari	2	12	-	-	-	-
30. Harijan Vannan	3	11	-	-	-	-
31. Ottan	4	13	-	-	-	-
32. Domban	9	39	0.60	0.04	-	-
33. Sukkillian	5	22	-	-	-	-
Sub-total	443	1,771	181.65	10.96	274.14	17.20
TOTAL	874	3,496	1,656.63	100.00	1,593.54	100.00

Note: The "forward", "backward", and "scheduled" castes are official categories employed by the Government of Tamil Nadu.

of the total agricultural land owned by the villagers. In contrast, the scheduled caste *(harijan)* groups consist mostly of landless and marginal landholding families, although their numbers in the population account for more than half of the village total. Between the above two categories of caste groups, there is still another category of caste groups, including those of the service caste, and there are great variations in their landholdings.

Among the caste groups, eight (*Reddiars*, *Udaiyars*, *Gounders*, Muslims, *Pallans*, *Paryans*, Catholic *Pallans*, and Catholic *Parayans*) have a considerable number of households and population and have distinct individual residential areas which originally centered around the *Reddiars* (fig. 6). Some other minor caste members like the *Chettiars* live within the above major caste groups without forming their own sizeable residential areas. Still another unique type is the *Nadar* caste, each household of which resides alone on the outskirts of the major residential areas or in the mid-fields and makes its living mostly as employed watchmen for the agricultural fields and gardens. Although there are some exceptions, the residential patterns in Peruvalanallur seem to reflect closely the segregation by caste which prevails in rural Indian communities.

CHAPTER III

LAND-USES AND THEIR ASSOCIATIONS

Seasonal Crops and Land-use Patterns

In terms of growing and harvesting of major crops (different types of paddy) the agricultural cycle in Tamil Nadu can be largely divided into three seasons: (1) June-September (during the southwest monsoon), (2) September-March (based on the northeast monsoon), and (3) March-June (the dry season). The first and second seasons are called respectively *kuruvai* and *thaladi* or *samba*, depending on the predominant cultivation pattern. More specifically, the classification of the *thaladi* paddy and *samba* paddy in the second season is based on whether or not the cultivation is associated with the *kuruvai* paddy cultivation in the previous season; if it is associated, it is called *thaladi*, and if not, it is called *samba*. It should be noted, however, that these seasons overlap each other rather than being sharply separated. Physical factors, including latitudinal differences and regional physiographic conditions, and managerial factors, including timely procurement of the labor forces for planting and harvesting and tractors and bullocks for ploughing and threshing, usually create an overlap of around one to two months between seasons.

Figures 7 and 8 and tables 4 and 5 show the spatial distribution of seasonal crops and their areas in Peruvalanallur in 1978-79 and 1979-80. As indicated in chapter I, agricultural lands in Tamil Nadu have been officially classified into two major categories, wet (lower) and dry (higher) lands. The former is usually sub-classified into "single" and "double" cropping lands based on whether one or two of the major crops can be cultivated on a particular piece of land.[1] It should be noted that there are some

---

1 . The above official classification of land has been made primarily for the revenue assessment of a particular parcel. According to the classification, land taxes are currently settled as follows:

1. In lower "wet land," land taxes for single and double cropping of paddy are respectively Rs.50/acre and Rs.80/acre; whilst for sugar cane and bananas they are Rs.90/acre.

2. In the irrigated areas under higher "dry land" where government sources of water are used, land taxes for single and double cropping of paddy are Rs.45/acre and Rs.60/acre respectively, and for sugar cane Rs.65/acre. Land taxes for both privately irrigated and for only rain-fed lands are as low as Rs.3.75/acre.

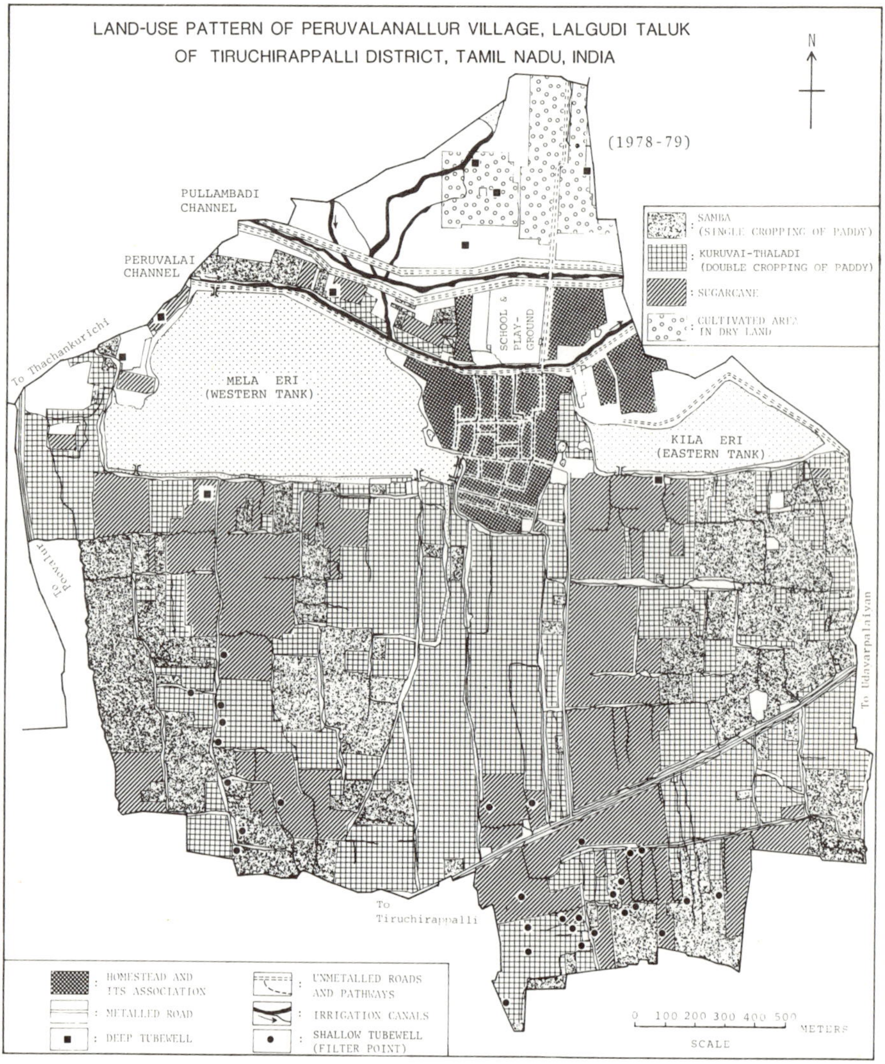
LAND-USE PATTERN OF PERUVALANALLUR VILLAGE, LALGUDI TALUK
OF TIRUCHIRAPPALLI DISTRICT, TAMIL NADU, INDIA
N
(1978-79)
PULLAMBADI CHANNEL
PERUVALAI CHANNEL
To Thachankurichi
MELA ERI (WESTERN TANK)
SCHOOL & PLAY-GROUND
KILA ERI (EASTERN TANK)
To Poovalur
To Udavarpalaiyan
To Tiruchirappalli
SAMBA (SINGLE CROPPING OF PADDY)
KURUVAI-THALADI (DOUBLE CROPPING OF PADDY)
SUGARCANE
CULTIVATED AREA IN DRY LAND
HOMESTEAD AND ITS ASSOCIATION
METALLED ROAD
DEEP TUBEWELL
UNMETALLED ROADS AND PATHWAYS
IRRIGATION CANALS
SHALLOW TUBEWELL (FILTER POINT)
0 100 200 300 400 500 METERS
SCALE

Fig. 7

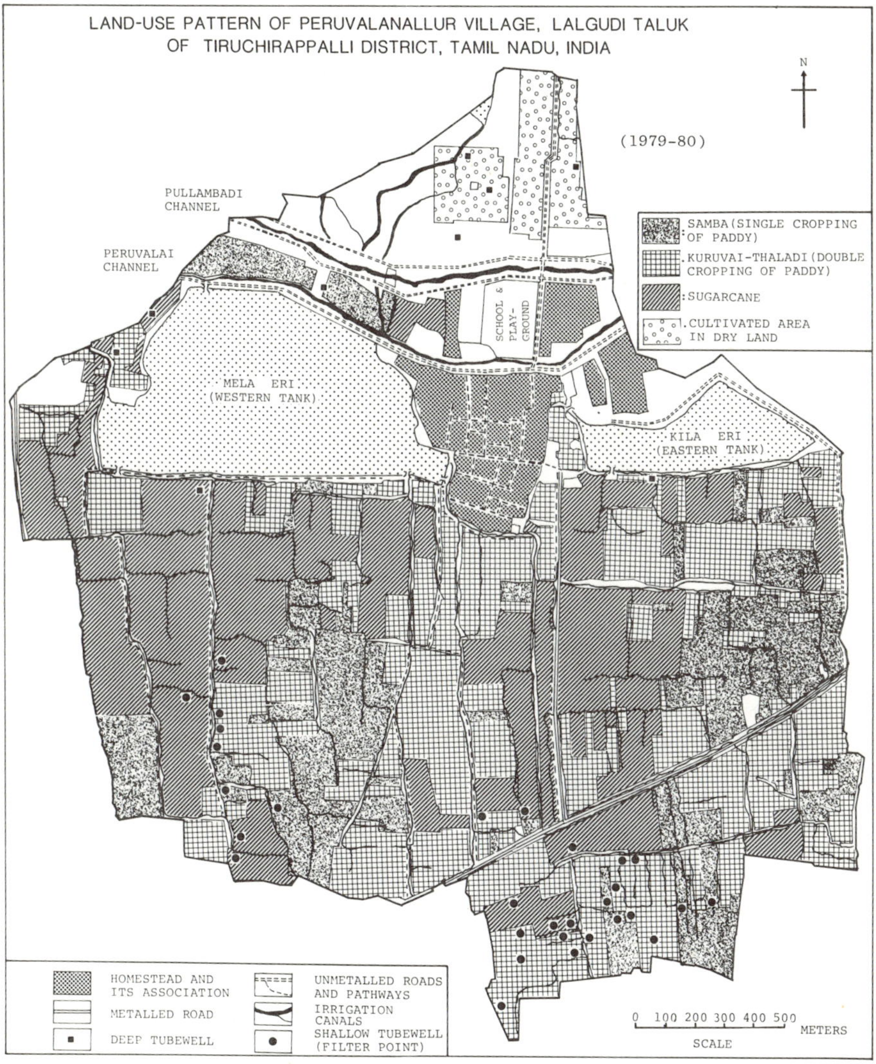
LAND-USE PATTERN OF PERUVALANALLUR VILLAGE, LALGUDI TALUK
OF TIRUCHIRAPPALLI DISTRICT, TAMIL NADU, INDIA
N
(1979-80)
PULLAMBADI CHANNEL
PERUVALAI CHANNEL
SCHOOL & PLAY-GROUND
MELA ERI (WESTERN TANK)
KILA ERI (EASTERN TANK)
SAMBA (SINGLE CROPPING OF PADDY)
KURUVAI-THALADI (DOUBLE CROPPING OF PADDY)
SUGARCANE
CULTIVATED AREA IN DRY LAND
HOMESTEAD AND ITS ASSOCIATION
METALLED ROAD
DEEP TUBEWELL
UNMETALLED ROADS AND PATHWAYS
IRRIGATION CANALS
SHALLOW TUBEWELL (FILTER POINT)
0 100 200 300 400 500
METERS
SCALE

Fig. 8

TABLE 4

LAND-USES OF PERUVALANALLUR VILLAGE OF LALGUDI TALUK, TIRUCHIRAPPALLI DISTRICT, TAMIL NADU, INDIA

(1978-79)

Cropping Seasons	Wet (Lower) Lands			Dry (Higher) Lands		
	Land-uses	Area (in Acre)	%[1]	Land-uses	Area (in Acre)	%
1st Season (June-Sept.)	Kuruvai	346.91	42.03	Kuruvai	7.40	6.06
	Sugarcane[2]	224.32	27.18	Sugarcane	10.20	8.35
	Vacant	254.15	30.79	Fruit Trees[3]	18.59	15.22
				Vacant	33.51	27.44
				Uncultivated	52.43	42.93
	Total	825.38	100.00	Total	122.13	100.00
2nd Season (Sept.-Mar.)	Thaladi	346.91	42.03	Thaladi	7.40	6.06
	Samba	254.15	30.79	Samba	11.66	9.55
	Sugarcane	224.32	27.18	Sugarcane	10.20	8.35
				Groundnut	12.12	9.92
				Red-gram	4.97	4.07
				Field Beans	0.50	0.41
				Maize	1.00	0.82
				Chilies	3.24	2.65
				Fruit Trees	18.59	15.22
				Others	0.02	0.02
				Uncultivated	52.43	42.93
	Total	825.38	100.00	Total	122.13	100.00
3rd Season (Mar.-June)	Sugarcane	224.32	27.18	Sugarcane	10.20	8.35
	Black-gram	10.29	1.25	Fruit Trees	18.59	15.22
	Others	0.29	0.04	Vacant	40.91	33.50
	Vacant	590.48	71.53	Uncultivated	52.43	42.93
	Total	825.38	100.00	Total	122.13	100.00

Source: The *Adangal* for the 1978-79 crop year.

Notes : 1. Percentages of each crop area are based on the net area for each season.

2. Sugarcane, which appears in the first, second, and third seasons in this table, is one crop.

3. Fruit Trees (18.59 acres) consist of lime (10.45 acres), palm (4.48), coconut palm (3.12), mango (0.48), and tamarind (0.06).

TABLE 5

LAND-USES OF PERUVALANALLUR VILLAGE OF LALGUDI TALUK, TIRUCHIRAPPALLI DISTRICT, TAMIL NADU, INDIA

(1979-80)

Cropping Seasons	Wet (Lower) Land			Dry (Higher) Land		
	Land-uses	Area (in Acre)	%[1]	Land-uses	Area (in Acre)	%
1st Season (June-Sept.)	Kuruvai	325.87	39.57	Kuruvai	6.84	5.60
	Sugarcane[2]	324.83	39.44	Sugarcane	5.62	4.60
	Others	1.52	0.18	Fruit Trees	13.33	10.92
	Vacant	171.35	20.18	Vacant	39.61	32.43
				Uncultivated	56.73	46.45
	Total	823.57	100.00	Total	122.13	100.00
2nd Season (Sept.-Mar.)	Thaladi	325.87	39.57	Thaladi	6.84	5.60
	Samba	171.35	20.81	Samba	15.20	12.45
	Sugarcane	324.83	39.44	Sugarcane	5.62	4.60
				Groundnut	11.83	9.69
				Red Gram	6.90	5.65
				Field Beans	0.59	0.48
				Maize	2.50	2.05
				Black Gram	2.18	1.78
				Ragi	0.41	0.34
				Fruit Trees	13.33	10.91
				Uncultivated	56.73	46.45
	Total	823.57	100.00	Total	122.13	100.00
3rd Season (Mar.-June)	Sugarcane	324.83	39.44	Sugarcane	5.62	4.60
	Others	1.52	0.18	Fruit Trees	13.33	10.92
	Vacant	497.22	100.00	Vacant	46.45	38.03
				Uncultivated	56.73	46.45
	Total	823.57	100.00	Total	122.13	100.00

Source: The Adangal for the 1979-80 crop year.

Notes: 1. Percentages of each crop area are based on the net area for each season.

2. Sugarcane, which appears in the first, second, and third seasons in this table, is one crop.

irrigated areas in the higher land and that they have still been officially included within the category of "dry land." As in the case of lower wet land, the irrigated areas in "dry land" are officially sub-classified into "single" and "double" cropping lands. Thus, our arrangement of statistics follows to this convention as shown in tables 4 and 5.

As indicated by table 2, in 1978-79 of the 947 acres of agricultural land in Peruvalanallur Village, 825 acres, or 87 percent, are classed as wet lands and 122 acres, or 13 percent, as dry lands.[2]

The wet lands are divided into three classes: (1) double-cropped rice land (*kuruvai* in the first season, *thaladi* in the second season, and vacant in the third season), 347 acres or just over 36 percent of the total agricultural land; (2) single-cropped rice land (vacant in the first season, *samba* in the second season, and vacant in the third season), 254 acres or 27 percent of the agricultural land; (3) sugar cane (which grows through all three seasons), 224 acres, or 24 percent of agricultural land. For the dry lands the largest categories are: (1) uncultivated, 52 acres, or 6 percent of all agricultural land; (2) fruit trees, 19 acres, or 2 percent; (3) sugar cane, 10 acres or 1 percent (all three of these categories occupy land in all three seasons); (4) single-cropped rice land, 12 acres or about 1 percent; (5) double-cropped rice land, 7 acres or less than 1 percent; and (6) other crops about 2 percent of the land (groundnuts 12 acres, red gram 5 acres, chillies 3 acres, and maize 1 acre). The small acreages of single-cropped land, double-cropped land, and sugar cane within the "dry lands" represent, of course, small irrigated patches amid the generally dry higher land in the northern part of the village.

As shown on figures 7 and 8, the wet lands lie south of and under the two tanks, which provide most of the irrigation water supplemented by shallow tubewells. The small area of dry land lies on higher ground in the northern part of the village. Within it small patches are irrigated by deep tubewells. The main residential area (homestead and its association) lies between the eastern and western water tanks and also between the dry lands to the north and

---

2. Figures in the text often are rounded to nearest whole acres or percent.

the wet lands to the south. A very small band of wet land lies north of the western tank.

Differences in land-use from one year to the next may be studied by comparing figure 7 for 1978-79 with figure 8 for 1979-80. The boundaries between dry lands and wet lands and for the residential areas are highly stable, but within the wet lands there is considerable variation from year to year in which lands are used for sugar cane, for double-cropping, or for single-cropping. *Kuruvai* paddy is transplanted during June and July. As the southwestern monsoon has never provided enough rain water directly to the study area for the early stage of cultivation, it is very dependent upon tank irrigation, water for which is partly provided by the Peruvalai channel, and upon direct irrigation from the channel. Therefore, one of the most important factors for a successful *kuruvai* cultivation is whether enough irrigation water can be provided from the channel, which is controlled at the Upper Anicut (barrage)[3] (located some 20 km west-northwest of Tiruchy) by the Public Works Department (PWD) under the Tamil Nadu Government (fig. 3). As the upper areas along the irrigation channel usually receive water earlier than the lower areas, the agricultural cycle in the former generally starts earlier than in the latter.[4]

ADT 31, which is regarded as an early maturing variety of paddy (rice plant), taking 105 days on an average from planting to harvest, is exclusively employed for the *kuruvai* paddy. The paddy in and around the study area is harvested mostly during September and October. However, it should be noted that the harvesting period overlaps partly with the northeastern monsoon (October-December). For this reason, the farmers desire to complete the entire harvest as quickly as possible. Early harvesting of *kuruvai* paddy is important in many ways: (1) the entire harvesting itself, including cutting, carrying, and threshing, will not be damaged by the rains; (2) the early paddy is of higher quality and, accordingly, its market price is higher than that for the same paddy harvested at a later period;[5] and (3) it provides for the early preparation and planting of the next crop, *thaladi* paddy, in the same fields.

---

3. Anicut from the Tamil *ani kattu* is a dam in a river for diverting water to a canal channel for irrigation.

4. For example, the harvesting of *kuruvai* paddy in the upper area near Lalgudi town located some 6 km southwest from the study area started 10-15 days earlier than in the study area.

5. The market price in 1979 for unhusked *kuruvai* paddy harvested before early October was Rs.60 per bag (1 bag is about 60 kg), but that of the same paddy harvested after November was Rs.50 per bag.

Even if farmers have successfully harvested the *kuruvai* paddy in time before the monsoon rains start, the following work for grain-drying cannot be free from the negative effects of the monsoon. In fact, grain-drying of *kuruvai* in the streets and on the housetops[6] in this season is frequently interrupted by sudden showers. It is in this season that almost all available space in farmhouses is intensively used for drying the paddy. In fact in the villages people commonly sleep on the floor surrounded by paddy grains.[7] In spite of the farmers' great efforts, the *kuruvai* is generally regarded as of lower quality[8] and its market price on the average is usually lower than those of other types of paddy (table 6).

Thaladi paddy is transplanted mostly during September and October and harvested from February through March. As indicated above, the timing of the planting of *thaladi* is dependent upon the *kuruvai* paddy harvest, since it is planted in the same fields. The variety employed for *thaladi* paddy is exclusively IR-20, a medium maturing variety rice plant taking 135 days on an average to harvest. Early planting of the crop is preferred in some fields where irrigation water is seldom provided from late January through March, and thus it can be harvested before facing a scarcity of water. However, if irrigation water can be supplied in the later stage of the cultivation, late harvesting in March in general provides a higher yield and better quality.[9] For this reason some innovative farmers have recently introduced shallow tubewells with diesel-powered pumps,[10] which are distributed mostly in the southwestern and southern sections of the study area, for their water-needy fields (figs. 7 and 8).

The *samba* area has been officially treated as a "single-cropping area" in the classification of land because there is no other major crop cultivated on it during the year. The *samba* season varies from 135 to 180 days depending not only upon the timing of planting, but also upon the paddy varieties employed. The

---

6. Each housetop usually has a "hole" with a diameter of 15-20 cm which leads to a store room underneath. This is a device for quick storing of grains and saving labor as well.

7. Most villagers in Tamil Nadu sleep on the floor with or without a bed sheet, not in a bed at all times.

8. Some villagers belonging to higher castes openly state, "We don't eat *kuruvai,* only *thaladi* or *samba. Kuruvai* is for *Harijans*."

9. The author assumes that the climatic factors, especially the temperature which increases gradually from January to April, might be responsible for this fact.

10. The sites of the shallow tubewells are officially called "filter points."

TABLE 6

YIELDS, COSTS, AND BENEFITS FOR SEASONAL CROPS IN PERUVALANALLUR VILLAGE,
LALGUDI TALUK OF TIRUCHIRAPPALLI DISTRICT, TAMIL NADU, INDIA

(1979-80)

Seasonal Crops	Yields (per Acre)		Price per Unit	Total Yields (per Acre)	Costs (per Acre)					Total Costs (per Acre)	"Net Benefit" (per Acre)
					Laborers						
	Grains (unhusked)	Straw			Harvesting	Others	Plowing	Ferti-lizers	Others		
Kuruvai paddy	33.0 bags	Rs. 100	Rs. 55 per bag	Rs. 1,965	3.0 bags (Rs. 165)	Rs. 150	Rs. 90	Rs. 400	Rs. 70	Rs. 875	Rs. 1,090
Thaladi Paddy	27.5 "	Rs. 150	Rs. 70 "	Rs. 2,075	2.5 bags (Rs. 175)	Rs. 150	Rs. 90	Rs. 400	Rs. 70	Rs. 885	Rs. 1,190
Samba Paddy	22.0 "	Rs. 150	Rs. 75 "	Rs. 1,800	2.0 bags (Rs. 150)	Rs. 150	Rs. 90	Rs. 400	Rs. 70	Rs. 860	Rs. 960
Sugarcane	50.0 ton (metric)	-	Rs. 132 per ton (at Factory)	Rs. 6,600	Rs. 750	Rs. 300	Rs. 90	Rs. 1,000	Rs. 1,135	Rs. 3,275	Rs. 3,325
Gram	5.0 bags	-	Rs. 110 per bag	Rs. 550	Rs. 40	-	-	-	Rs. 10	Rs. 50	Rs. 500

Types of Crop Associations	Annual "Net Benefits" (per Acre)
1. Samba paddy only	Rs. 960
2. Samba paddy→Gram	Rs. 1,460
3. Kuruvai paddy→Thaladi paddy	Rs. 2,280
4. Kuruvai paddy→Thaladi paddy→Gram	Rs. 2,780
5. Sugarcane only	Rs. 3,325
6. Sugarcane→Gram	Rs. 3,825

Note:
1. Individual paddy prices were those at the studied village.
2. The cost of each harvesting is based on the amount of each crop's yield, because the laborer gets a fixed ratio of his total harvest.
3. Transportation cost for sugarcane is Rs. 5/ton per kilometer. As the road distance between the main gate of the village and the nearest sugar factory is about 4 km., the transportation change for 50 tons (the yields per acre) of sugarcane is Rs. 1,000 which is included under "Others" in the table.
4. One bag is about 60 kg.

varieties of rice plants employed are either IR-20 or *ponni,* which are regarded respectively as medium maturing (135 days on an average) and late maturing (150 days). *Samba* paddy is generally transplanted from August through October, *ponni* (a late maturing variety) being employed for early planting; whereas IR-20 (a medium maturing variety) is used for late planting. Because of these selections of varieties, the harvesting period is almost the same as for *thaladi.* Nonetheless, some fields are planted earlier with the *ponni* variety so that they can be harvested before early January, since this variety is usually used as one of the important cooking ingredients for the *Pongal* Festival observed on January 14 and 15 every year.

Besides the above-mentioned different types of paddy, sugar cane and various types of gram are important crops in the wet lands of Peruvalanallur village. Since cane takes almost one year to harvest, its area is regarded officially as a "double-cropping area" like that of the *kuruvai* and *thaladi* paddy fields. Sugar cane has great variations in its planting season (from December to May of the next year) and is harvested one year after the planting. In Peruvalanallur and the surrounding villages, sugar cane is planted mostly from January through March and thus harvested in the same period the following year. As this harvesting period corresponds to those of *thaladi* and *samba,* it is the busiest season of the year, followed by the period between September and October when the harvesting of *kuruvai* paddy and the plantings of *thaladi* and *samba* take place. Once sugar cane is planted, it usually continues to be cultivated for two consecutive years and then is replaced by paddy, although there are some cases in which it is continued for three years. This is due to the fact that the yields of sugar cane diminish gradually from the first year's harvest to the second and the third. Of course, whether or not sugar cane (or paddy) should be cultivated in particular fields in a year depends upon the results of the farmers' "benefit-cost analysis." Yet, there has been an increasing interest among villagers in sugar cane cultivation for which the increasing market price of processed sugar in recent years has been mostly responsible.

Various types of gram and other field crops have been extensively cultivated in the wet lands of the study area during the earlier part of the dry season, although only a few acres were recorded in the *Adangal*s for 1978-79 and 1979-80. As far as the extent of the cultivated area is concerned, black gram is the most important among various types of gram in the wet land. The gram takes 65 days on an average to harvest. The fields for gram are the same ones which have been used for the *thaladi* paddy or *samba* paddy in the previous (second) season, although its association with *thaladi* is more commonly observed in Peruvalanallur. Gram is usually sown in the fields broadcasting, where *thaladi* or *samba* is still standing but is about ready to be harvested in 7-10 days. This practice is regarded as the farmers' ingenious idea for the effective utilization of soil moisture for the crops. The timing of this sowing and, accordingly, of its harvesting depends entirely upon that of the paddy harvests. Thus, gram in some fields is harvested even in mid-March. After the early harvests of the gram, some of the same fields are immediately used for gingelly, which takes 70 days on an average to harvest. Of course, this practice is only possible in specific fields where adequate soil moisture is available. In fact, gingelly cultivation is observed in the fields neighboring the sugar cane areas which receive irrigation water occasionally during the dry season.

Gram and other field crops in the wet lands do not require much cost and care for their cultivation, according to the villagers, and still give appreciable returns (table 6). Thus, at this stage, it should be pointed out that the traditional categories of the "single-cropping" and "double-cropping" areas include such field crops to a lesser or greater extent, and that this paper itself follows this convention.

Agricultural Lands and Irrigation Systems

The available historical documents, the *First Settlement Register* (1864), the *First Re-settlement Register* (1898), and the *Second Re-settlement Register* (1927), indicate that there has been a piecemeal development of the irrigated area in Peruvalanallur during the second half of the 19th century and the first quarter of this century (cf. figs. 9, 10, and 11 and table 7). Of the above documents, the last one provides the most detailed information on the related aspects of irrigation.

Over this period of somewhat more than a century (1864-1980), the total agricultural land of the village remained relatively stable, changing only from 1,049 acres in 1864 to 1,043 in 1927 but then declining by nearly ten percent from 1,043 acres to 946 acres in 1979-80 (table 7). This decline reflects partly the conversion of agricultural land to other uses such as roads, pathways, canals, residences, and tanks or other water bodies.

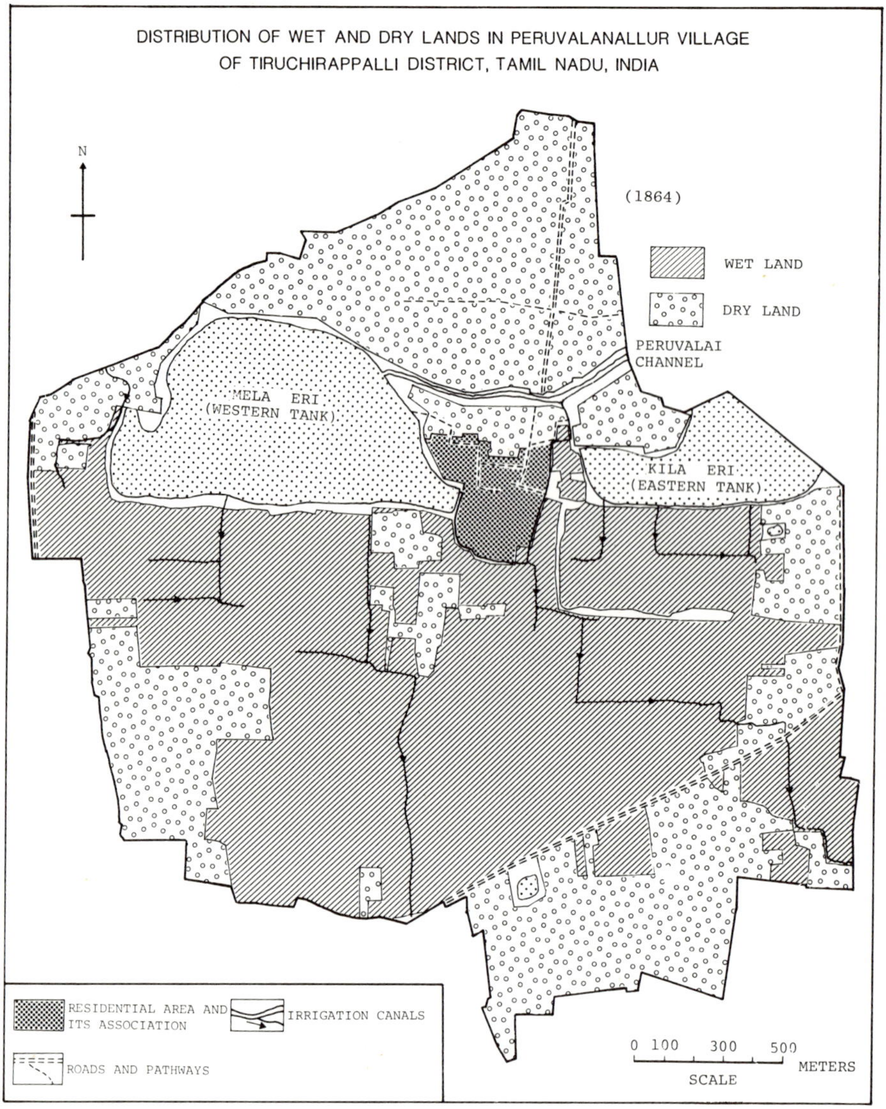
DISTRIBUTION OF WET AND DRY LANDS IN PERUVALANALLUR VILLAGE
OF TIRUCHIRAPPALLI DISTRICT, TAMIL NADU, INDIA
N
(1864)
WET LAND
DRY LAND
PERUVALAI
CHANNEL
MELA ERI
(WESTERN TANK)
KILA ERI
(EASTERN TANK)
RESIDENTIAL AREA AND
ITS ASSOCIATION
IRRIGATION CANALS
ROADS AND PATHWAYS
0 100 300 500
METERS
SCALE

Fig. 9

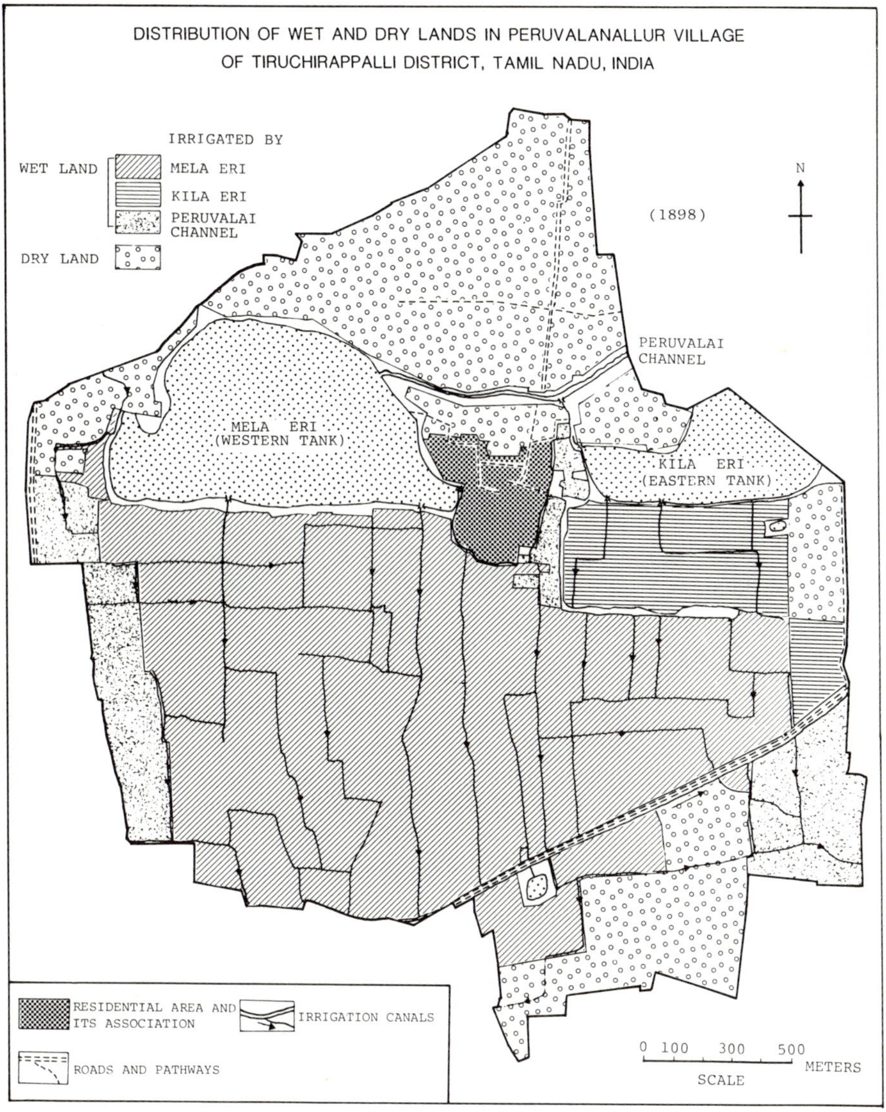
DISTRIBUTION OF WET AND DRY LANDS IN PERUVALANALLUR VILLAGE
OF TIRUCHIRAPPALLI DISTRICT, TAMIL NADU, INDIA
IRRIGATED BY
WET LAND
MELA ERI
KILA ERI
PERUVALAI CHANNEL
DRY LAND
(1898)
N
PERUVALAI CHANNEL
MELA ERI (WESTERN TANK)
KILA ERI (EASTERN TANK)
RESIDENTIAL AREA AND ITS ASSOCIATION
IRRIGATION CANALS
ROADS AND PATHWAYS
0 100 300 500
METERS
SCALE

Fig. 10

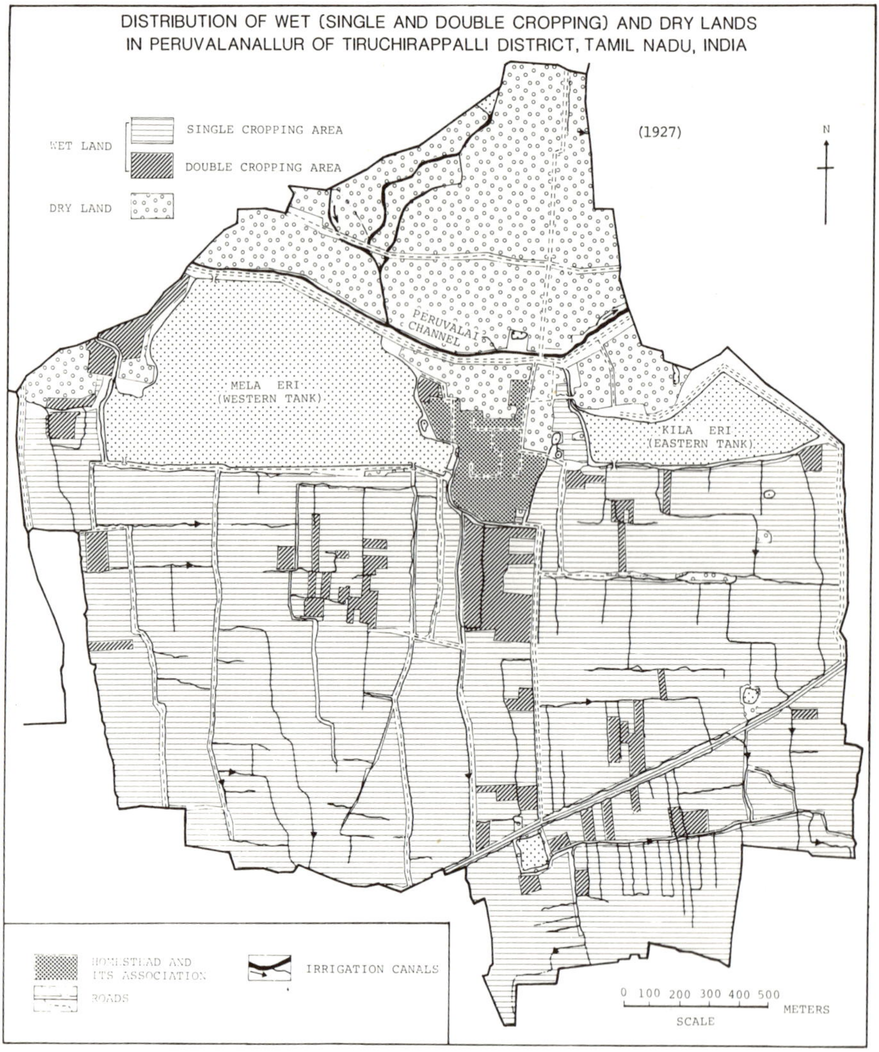
DISTRIBUTION OF WET (SINGLE AND DOUBLE CROPPING) AND DRY LANDS
IN PERUVALANALLUR OF TIRUCHIRAPPALLI DISTRICT, TAMIL NADU, INDIA
(1927)
N
WET LAND
SINGLE CROPPING AREA
DOUBLE CROPPING AREA
DRY LAND
PERUVALAI CHANNEL
MELA ERI
(WESTERN TANK)
KILA ERI
(EASTERN TANK)
HOMESTEAD AND ITS ASSOCIATION
ROADS
IRRIGATION CANALS
0 100 200 300 400 500
METERS
SCALE

Fig. 11

TABLE 7

DEVELOPMENT OF AGRICULTURAL LANDS: DRY TO SINGLE CROPPING WET LANDS; SINGLE CROPPING WET TO DOUBLE CROPPING WET LANDS IN PERUVALANALLUR VILLAGE OF LALGUDI TALUK, TIRUCHIRAPPALLI DISTRICT TAMIL NADU, INDIA (1864 - 1980)

(Unit: in acre)

	1864	1898	1927	1978-79	1979-80
Dry Lands	480.53	330.09	209.61	122.13	122.13
Wet Lands					
Single Cropping	568.03	718.47	770.67	254.15	171.35
Double Cropping	-	-	62.82	571.23	652.22
Sub-Total	(568.03)	(718.47)	(833.49)	(825.38)	(823.57)
Total	1,048.56	1,048.56	1,043.10	947.51	945.70

Sources:
1. The First Settlement Register, 1864
2. The First Re-settlement Register, 1898
3. The Second Re-settlement Register, 1927
4. The Village Map for Peruvalanallur, 1893
5. The Village Map for Peruvalanallur, 1927
6. The Field Map Books for Peruvalanallur, (current)
7. The Adangals (the Annual Land-use Records) for 1978-79 and 1979-80

The major change over this period has been an intensification of land use. The relatively low-production dry lands have declined by three-quarters from 481 acres in 1864 to 122 acres in 1980, whereas the high-production wet lands have increased from 568 acres to 824 acres. This change occurred during the early part of the period between 1864 and 1927 as the acreage in wet lands has actually declined slightly since 1927. Another expression of intensification has occurred within the wet lands. In 1864, wet lands were all devoted to single-cropping (568 acres), which expanded to 771 acres in 1927 but then dropped sharply to 171 acres in 1980. Large areas were shifted to the more intensive double-cropping, absent in 1864 and 1898, rising to 63 acres in 1927, and then expanding very rapidly to 652 acres in 1980 to occupy 69 percent of all agricultural land in the village. The location and expansion of wet and dry land in Peruvalanallur Village over the period 1864 to 1980 is depicted on figures 9 to 13.

In 1864, the area in wet lands formed a nearly solid central block immediately south of the two tanks and the residential area. It was surrounded by dry lands, nearly as large in total area (table 7), with large blocks in the higher land north of the residential area, other blocks on the southwest and southeast beyond the wet lands and more remote from the tanks and the residential area, and a few small patches within the wet lands (fig. 9).

By 1898, the internal patches, the southwest block and part of the southeast block had been converted from dry land to wet land (fig. 10). Available data also permit the recording on this map of the source of water for the wet lands. Most of this land was irrigated by water from the western tank *(mela eri)*, but a small area just south of the eastern tank *(kila eri)* received water from that tank. Peripheral strips along the western and eastern edges of the southern wet lands received water directly from the Peruvalai channel.

By 1927, the remaining village land south of the water tanks had been converted from dry land to wet land so that the land south of the residential area now formed a solid continuous block of wet land, whereas the land north of the residential area remained a solid block of dry land. With the coming of supplementary water to extend the irrigation season, patches of double-cropping appeared for the first time on the 1927 map, mainly in the area west of and immediately adjacent to the western tank but with scattered patches elsewhere to the south and southeast.

By 1978-79, much of the wet land had been converted to double-cropping, but areas of single-cropping remained particularly along the western edges of the village wet lands. The rapidity of transformation from single-cropping to double-cropping is revealed by the comparison of 1979-80 with 1978-79 (figs. 12 and 13 and table 7). During this year, the amount of single-cropped land decreased from 254 to 171 acres as such land particularly along the western edge of the wet lands was converted to double-cropping.

There still remains some potential for further conversion from single-cropping to double-cropping as indicated by figure 14, but some wet lands appear likely to remain single-cropping areas because of difficulty of supplying extra water to extend the cropping seasons. Almost all of the presently existing irrigation networks in the wet lands of Peruvalanallur had been shaped before 1927 and account for the rise in wet lands between 1864 and 1927. It should be noted that the Peruvalai channel appeared in the index map of the *First Re-settlement Register* (1898). The major contribution to the remarkable expansion of the double-cropping area since 1927 must have been due to the Mettur Dam of Salem District, located some 178 km northwest of the study area, the construction of which was completed in 1934, and which made possible the supply of water over a longer period.

The seasonal crops in the Peruvalanallur village show some characteristic distributions (figs. 7 and 8). Roughly speaking, the double-cropping areas of *kuruvai* and *thaladi* paddy, and sugar cane[11] were distributed over the major parts of the wet lands of the study area, whereas the single-cropping areas of *samba* paddy were found in the western and eastern peripheries. The reason why the particular fields have been assigned to the *samba* paddy cultivation is obviously a poor supply of irrigation water except during the second season, that of the relatively abundant rains of the northeast monsoon. As mentioned earlier, the two tanks, the *mela* and *kila eris*, and the Peruvalai channel are major sources of irrigation water for the study area. However, each of the tanks keeps water at a maximum level only during the northeast monsoon period (October-December), and their water levels usually decrease gradually toward the dry season, although occasionally the water supply for the tanks has been provided during the dry season through the channel controlled by the Public Works Department. The tanks drain in different directions: the floor of the *mela eri* is lowest

---

11. As stated before, the sugar cane fields are officially treated as "double-cropping areas."

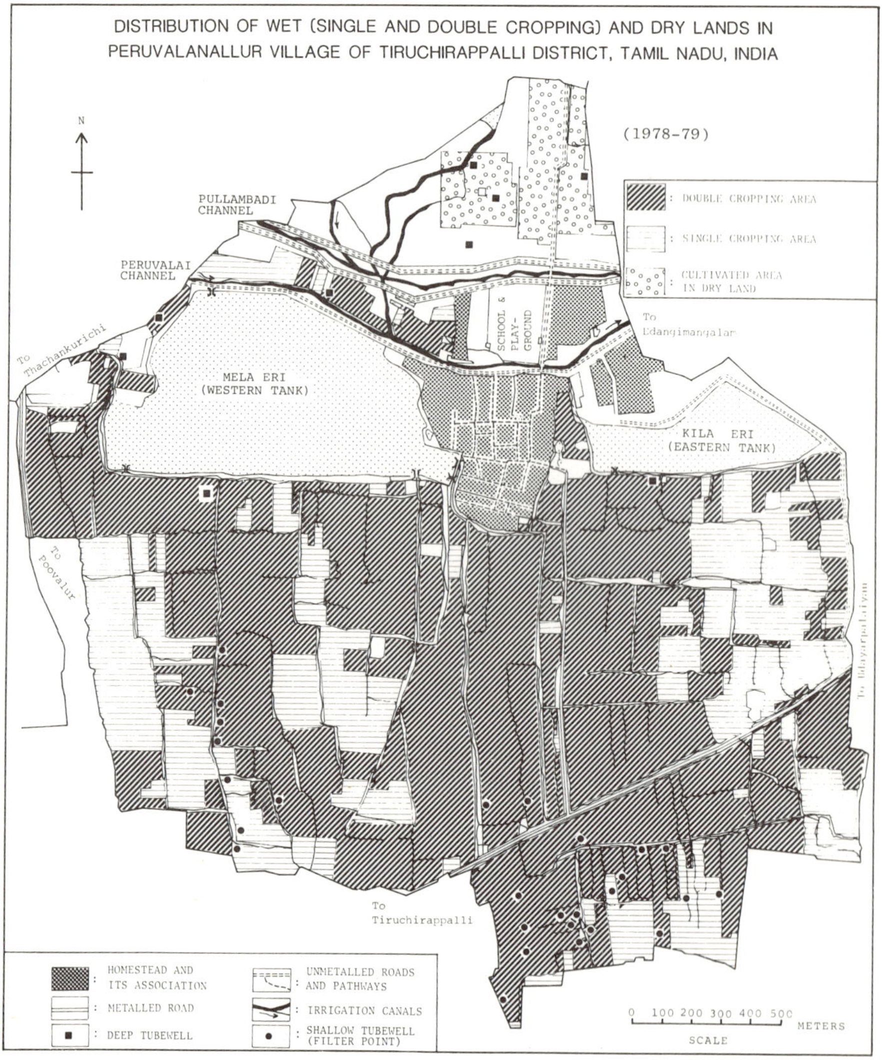
DISTRIBUTION OF WET (SINGLE AND DOUBLE CROPPING) AND DRY LANDS IN PERUVALANALLUR VILLAGE OF TIRUCHIRAPPALLI DISTRICT, TAMIL NADU, INDIA
N
(1978-79)
: DOUBLE CROPPING AREA
: SINGLE CROPPING AREA
: CULTIVATED AREA IN DRY LAND
PULLAMBADI CHANNEL
PERUVALAI CHANNEL
To Thachankurichi
MELA ERI (WESTERN TANK)
SCHOOL & PLAY-GROUND
To Edangimangalam
KILA ERI (EASTERN TANK)
To Poovalur
To Edayarpalaiyam
To Tiruchirappalli
: HOMESTEAD AND ITS ASSOCIATION
: METALLED ROAD
: DEEP TUBEWELL
: UNMETALLED ROADS AND PATHWAYS
: IRRIGATION CANALS
: SHALLOW TUBEWELL (FILTER POINT)
0 100 200 300 400 500
METERS
SCALE

Fig. 12

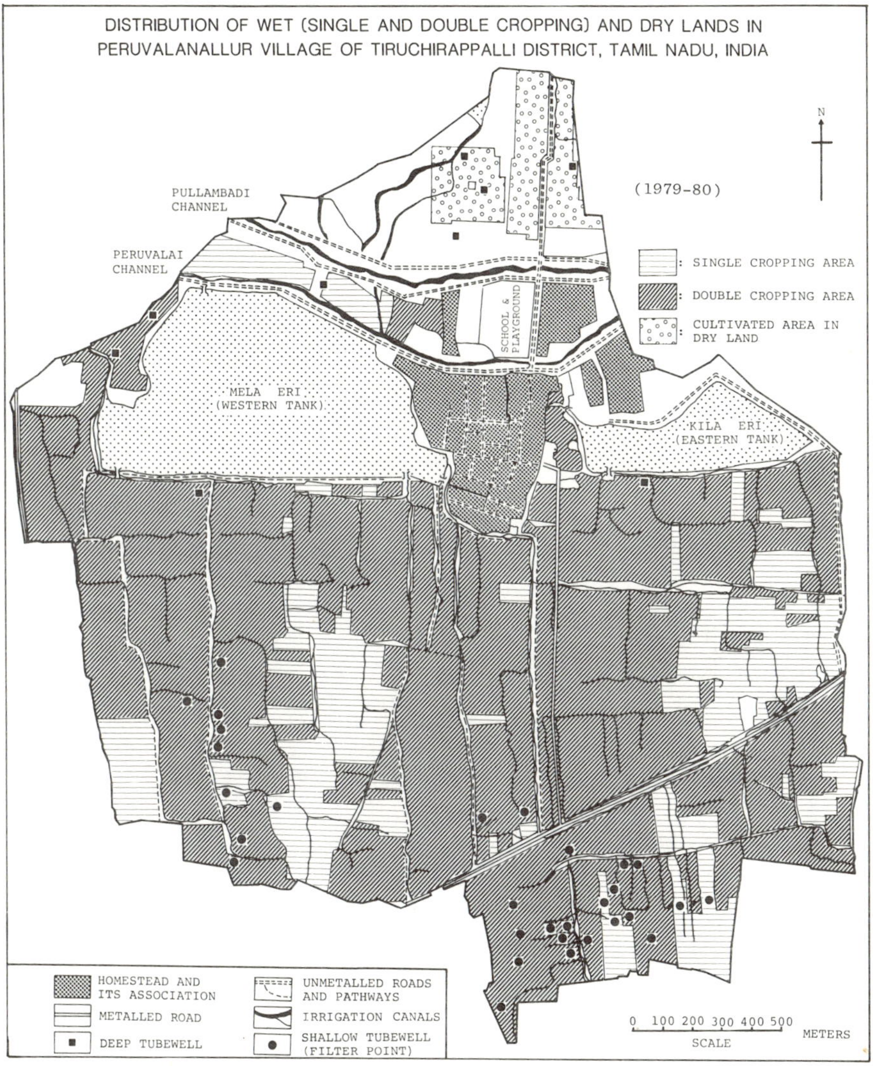
DISTRIBUTION OF WET (SINGLE AND DOUBLE CROPPING) AND DRY LANDS IN PERUVALANALLUR VILLAGE OF TIRUCHIRAPPALLI DISTRICT, TAMIL NADU, INDIA
N
(1979-80)
PULLAMBADI CHANNEL
PERUVALAI CHANNEL
SINGLE CROPPING AREA
DOUBLE CROPPING AREA
CULTIVATED AREA IN DRY LAND
SCHOOL & PLAYGROUND
MELA ERI (WESTERN TANK)
KILA ERI (EASTERN TANK)
HOMESTEAD AND ITS ASSOCIATION
METALLED ROAD
DEEP TUBEWELL
UNMETALLED ROADS AND PATHWAYS
IRRIGATION CANALS
SHALLOW TUBEWELL (FILTER POINT)
0 100 200 300 400 500
METERS
SCALE

Fig. 13

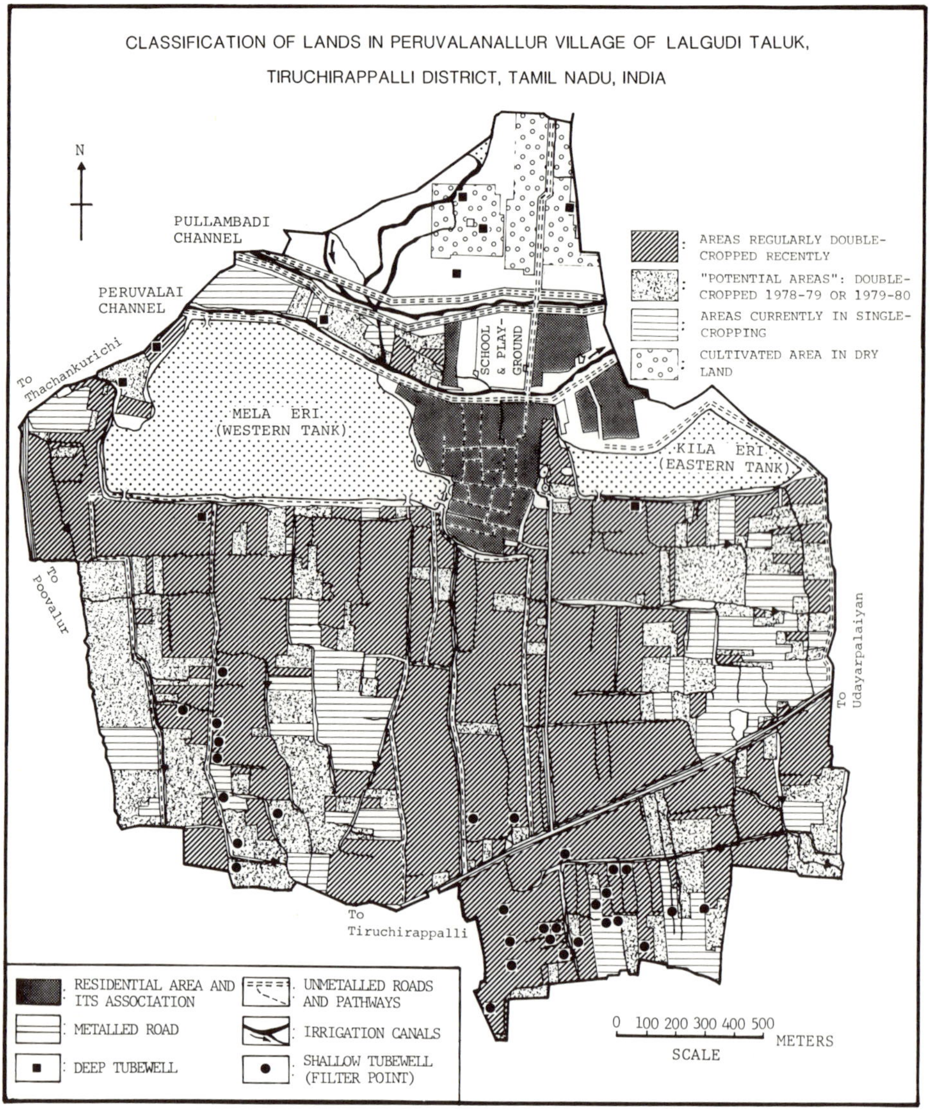
CLASSIFICATION OF LANDS IN PERUVALANALLUR VILLAGE OF LALGUDI TALUK,
TIRUCHIRAPPALLI DISTRICT, TAMIL NADU, INDIA
N
PULLAMBADI CHANNEL
PERUVALAI CHANNEL
To Thachankurichi
MELA ERI (WESTERN TANK)
SCHOOL & PLAY-GROUND
KILA ERI (EASTERN TANK)
To Poovalur
To Udayarpalaiyan
To Tiruchirappalli
: AREAS REGULARLY DOUBLE-CROPPED RECENTLY
: "POTENTIAL AREAS": DOUBLE-CROPPED 1978-79 OR 1979-80
: AREAS CURRENTLY IN SINGLE-CROPPING
: CULTIVATED AREA IN DRY LAND
: RESIDENTIAL AREA AND ITS ASSOCIATION
: METALLED ROAD
: DEEP TUBEWELL
: UNMETALLED ROADS AND PATHWAYS
: IRRIGATION CANALS
: SHALLOW TUBEWELL (FILTER POINT)
0 100 200 300 400 500
METERS
SCALE

Fig. 14

at its eastern corner and gradually rises toward the west; that of the *kila eri* is lowest at its western corner and highest at its eastern corner. Therefore, there is a great variation in their quantity of water and the period of water supply among the available field canals, depending upon the different locations of the water outlets of the tanks. In addition, it should be noted that there is also a great variation in the number of opening days (and hours) among the individual intakes of the Peruvalai channel. In this respect, the two intakes of the channel, located immediately east and west of the main residential area, and their field canals have together played an important role in the effective irrigation of the major part of the wet lands during the water-needy period, especially when the tanks' water levels are low or have virtually dried up.

Still other physical factors seem to be responsible for the single-cropping in particular areas, that is, land elevation and saline soil. Some of the fields in the single-cropping area are slightly higher in elevation and are distributed in a mosaic pattern throughout the wet lands of the village. The problems caused by the saline soil are generally acknowledged by villagers all over Lalgudi Taluk. In the study area, the most affected areas correspond closely to those of the single-cropping *samba* paddy, yields of which (15 bags per acre) account for about half that of *thaladi* paddy in the same (second) season. This problem also is related to the irrigation water supply, because the areas of saline soils need a lot of fresh water for a better crop yield.

Some factors responsible for the effective utilization of the irrigation facilities available in the study area should be pointed out. As already indicated, there is great variation in the size of landownership among the different caste groups. The *Reddiars*, for example, occupied in 1980 nearly 60 percent of the total land owned by the Peruvalanallur villagers. Most of their lands are in the wet area, although they are, to some extent, mingled with those owned by other caste groups of the same village as well as with those owned by other villagers. As the leading caste group and "guardian" of the other caste groups, the *Reddiar* community as a whole has made great efforts to get "more water" to the village. For this purpose, some of their leaders have kept in regular contact with the local office of the Public Works Department (PWD) at Tiruchy and, when necessary, with its head office at Madras. In fact, the major part of the common fund of the *Reddiar* community has been spent every year for this purpose.

It is interesting to note that the distribution of lands owned by outsiders overlap to a great extent the single-cropping areas in the western and eastern sections of the village (fig. 7 and fig. 19). This fact is at least partly responsible for the poor supply of irrigation water, since the outsiders have failed to form an organized body to obtain more irrigation water.

There are some 30 shallow tubewells (filter points) distributed in the fields (9 installations) adjacent to the western single-cropping area and in the southeastern corner (21 installations) of the Peruvalanallur village. Mainly because of the supplementary water by the shallow tubewells, it has been possible to cultivate double crops *(kuruvai* and *thaladi)* or sugar cane with great security. The tubewell owners, who are residents of Peruvalanallur except for one owner, usually have installed filter points in and around their own sizeable land clusters. However, it does not necessarily mean that the tubewell owner has a perfect cluster of land fit for the irrigation potential of his tubewell. Rather, it is more common that the tubewell owner's fields are interspersed with others' fields on a contract basis. The tubewell water is distributed mostly either by a field-to-field irrigation method or by using the existing field canals. When part of the tubewell owner's fields are located at some distance (20-30 meters) from his main land cluster, there is still another irrigation method: using a pipe which runs under the ground of intervening fields.

Apart from the wet land irrigation, there are four modes of irrigation available for dry lands: (1) *aetram-eravai* (slope-lifting by people), (2) *kavalai* (skin-bag-lifting by a pair of bullocks), (3) low-lift diesel pumps, and (4) deep tubewells with electric power, although tank irrigation is usually practiced in other "dry villages." The first two modes are being practiced in a narrow strip area in the northwestern corner immediately north of the Peruvalai channel, and the last two are employed in the northern section of the Peruvalanallur village, where there are four deep tubewells (figs. 7 and 8). The area employing these latter two modes of irrigation, however, totaled only 40.09 acres in 1979-80.

CHAPTER IV
LANDOWNERSHIP AND ITS IMPLICATIONS

The Distribution of Landownership

It has been observed that in South Indian rural communities the agricultural lands in a given village are owned mostly by its own village residents and the Hindu temples, but partly by some outside villagers. Tables 8 and 9, and figures 15 and 16 show the spatial distribution of inter-village landownership.

In Peruvalanallur in 1980, there were 947.51 acres of agricultural lands of which its own villagers and the Hindu temples (and other public bodies) owned 681.29 acres (72 percent) and 65.01 acres (7 percent) respectively.

The remaining 201.21 acres (21 percent) were owned by outsiders: 191.41 acres (20 percent) by those who lived in 30 different villages and towns and 9.80 acres (1 percent) by the Hindu temples in 4 different villages (table 8). About half of the land in Peruvalanallur village owned by inhabitants of other villages is owned by villagers in two villages just north and northwest of Peruvalanllur, by those in Thachankurichi (no. 34 in fig. 4 and in table 1), more than 62 acres, and by those in Kumulur (no. 37), nearly 36 acres (table 8, fig. 15).

But inhabitants in Peruvalanallur village owned much more land in other villages, a total of 975.34 acres, nearly five times as much land as owned by outsiders in Peruvalanallur, and about equal to the total agricultural land in Peruvalanallur village itself (table 9). These outside holdings by villagers of Peruvalanallur are widely scattered through 31 other villages, mainly with Lalgudi Taluk but including small holdings also outside this *taluk*. About half of these external holdings are in Kumulur village (no. 37), adjoining Peruvalanallur on the north with more than 449 acres (table 9 and fig. 16), about a fourth in Edangimangalam (no. 84), lying immediately on the east with more than 247 acres, and substantial amounts also in Sirumayangudi (no. 62), 58 acres, and

TABLE 8

VILLAGES AND TOWNS WHOSE PEOPLE OWNED LANDS IN PERUVALANALLUR VILLAGE OF LALGUDI TALUK, TIRUCHIRAPPALLI DISTRICT, TAMIL NADU, INDIA (1980)

Name of Villages	Area (in Acre)	Name of Villages and Towns	Area (in Acre)
(A) Within Lalgudi Taluk:		(B) Outside Lalgudi Taluk:	
1. Mannachanallur (#14)	0.66	25. Tiruchy (Tiruchy T.K.)	3.59
2. Thachankurichi (#34)	62.44	26. Srirangam (Tiruchy T.K.)	0.28
3. Reddimangudi (#35)	10.23	27. Kalpadi (Parambalur T.K.)	0.74
4. Kumulur (#37)	35.92	28. Thiruverambur (Tiruchy T.K.)	0.13
5. Vellanur (#38)	0.19	29. Gandinagar (Tiruchy T.K.)	0.13
6. Peruvalappur (#42)	15.40	30. Neiveli (South Arcot D.T.)	0.37
7. Kanakkiliyanallur (#43)	4.12		
8. Pullambadi (#45)	0.66		
9. Pudur (#49)	0.11	(B) Sub-total	5.24
10. Alambadi (#50)	0.31		
11. Kilanbil (#58)	0.34	(C) Temples:	
12. Pallapuram (#63)	0.50		
13. Sirumayangudi (#62)	1.11	31. Pullambadi (#45)	1.00
14. Poovalur (#64)	7.54	32. Poovalur (#64)	2.94
15. Manakkal (#65)	5.76	33. Siruganur (#91)	4.38
16. Lalgudi (#69-71)	4.43	34. Uttathur (#116)	1.48
17. Pudur Uthamanur (#73)	11.29		
18. Perakambi (#86)	4.63		
19. Neykulam (#89)	1.60	(C) Sub-total	9.80
20. Siruganur (#91)	7.24		
21. Sirugudi (#98)	3.09		
22. Gurudamangalam (#104)	1.41	Note: Each number in parentheses corresponds to the village number listed in Table 1.	
23. Kallakudi (#112)	4.43		
24. Tiruppattur (#117)	2.76		
(A) Sub-total	186.17	TOTAL (A)+(B)+(C)	201.21

LANDS IN THE OTHER VILLAGES OWNED BY PERUVALANALLUR VILLAGERS OF
LALGUDI TALUK, TIRUCHIRAPPALLI DISTRICT, TAMIL NADU, INDIA

(1980)

Name of Villages	Area (in Acre)
(A) Within Lalgudi Taluk:	
1. Valadi (#26)	2.00
2. Sirumarudur (#27)	8.61
3. Turaiyur (#28)	22.74
4. Kannanur (#29)	2.87
5. Marudur (#30)	3.77
6. Valavanur (#31)	14.54
7. Thachankurichi (#34)	1.05
8. Peddimangudi (#35)	2.00
9. Kumulur (#37)	449.49
10. Vellanur (#38)	1.90
11. Sangendi [Punjai] (#40)	9.86
12. Venkatachalapuram (#46)	0.60
13. Sirumayangudi (#62)	58.09
14. Poovalur (#64)	33.01
15. Thirumangalam (#72)	12.58
16. Neykuppai (#74)	4.10
17. Edangimangalam (#84)	247.53
18. Siruganur (#91)	54.40
19. Kallakudi (#121)	5.10
(A) Sub-total	934.24

Name of Villages and Towns	Area (in Acre)
(B) Outside Lalgudi Taluk:	
20. Perumalpalayan (Thuraiyur T.K., Tiruchy D.T.)	4.25
21. Pulivalam (Thuraiyur T.K., Tiruchy D.T.)	0.75
22. Kannambody (Thuraiyur T.K., Tiruchy D.T.)	2.00
23. Thimmor (Perambalur T.K., Tiruchy D.T.)	10.10
24. Vepampoondi (Perambalur T.K., Tiruchy D.T.)	2.00
25. Varagupadi (Perambalur T.K., Tiruchy D.T.)	1.00
26. Annamangalam (Perambalur T.K., Tiruchy D.T.)	1.50
27. Kovil Esanai (Ariyalur T.K., Tiruchy D.T.)	1.00
28. Govindampalayam (Attur T.K., Salem D.T.)	1.50
29. Melanariyappanur (South Arcot D.T.)	0.50
30. Kottamoolanur (Dharapuram T.K., Coimbatore D.T.)	15.50
31. Vannar Palayam	1.00
(B) Sub-total	41.10
TOTAL (A) + (B)	975.34

Note:
Each number in parentheses corresponds to the village number listed in Table 1.

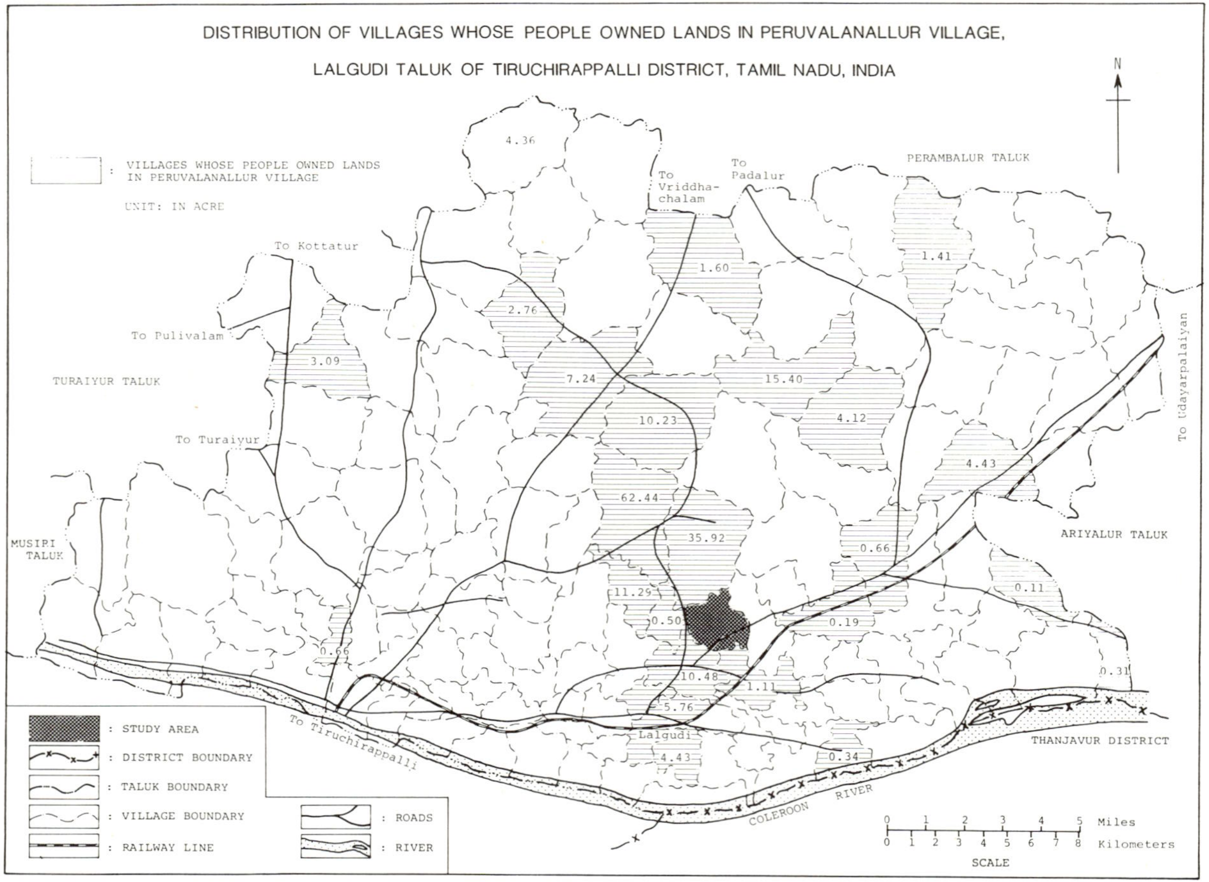
DISTRIBUTION OF VILLAGES WHOSE PEOPLE OWNED LANDS IN PERUVALANALLUR VILLAGE,
LALGUDI TALUK OF TIRUCHIRAPPALLI DISTRICT, TAMIL NADU, INDIA
N
: VILLAGES WHOSE PEOPLE OWNED LANDS IN PERUVALANALLUR VILLAGE
UNIT: IN ACRE
4.36
To Vriddha-chalam
To Padalur
PERAMBALUR TALUK
To Kottatur
1.60
1.41
2.76
To Pulivalam
3.09
TURAIYUR TALUK
7.24
15.40
10.23
4.12
To Turaiyur
4.43
To Udayarpalaiyan
62.44
35.92
0.66
ARIYALUR TALUK
MUSIRI TALUK
11.29
0.11
0.50
0.19
0.66
0.31
10.48
1.11
5.76
Lalgudi
4.43
0.34
To Tiruchirappalli
THANJAVUR DISTRICT
COLEROON RIVER
: STUDY AREA
: DISTRICT BOUNDARY
: TALUK BOUNDARY
: VILLAGE BOUNDARY
: RAILWAY LINE
: ROADS
: RIVER
0 1 2 3 4 5 Miles
0 1 2 3 4 5 6 7 8 Kilometers
SCALE

Fig. 15

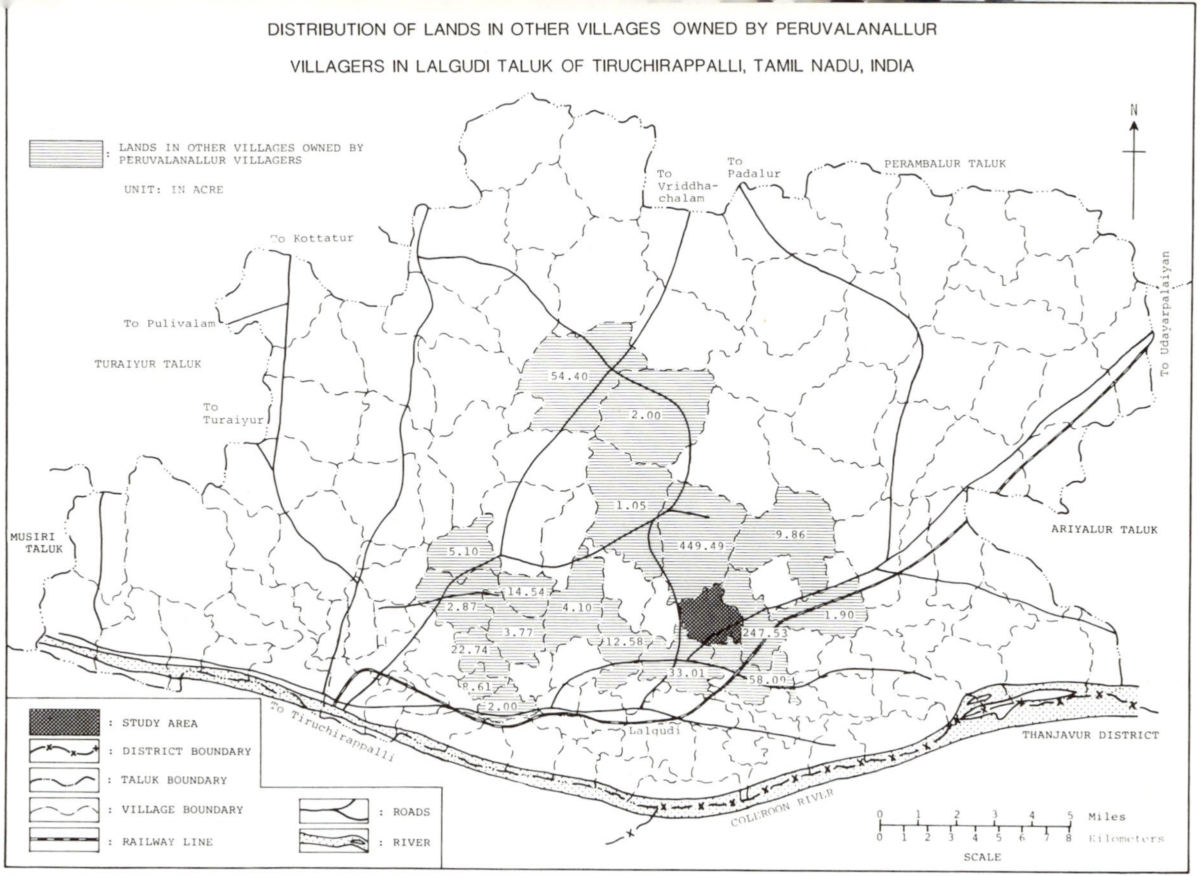
DISTRIBUTION OF LANDS IN OTHER VILLAGES OWNED BY PERUVALANALLUR
VILLAGERS IN LALGUDI TALUK OF TIRUCHIRAPPALLI, TAMIL NADU, INDIA
LANDS IN OTHER VILLAGES OWNED BY PERUVALANALLUR VILLAGERS
UNIT: IN ACRE
N
To Kottatur
To Pulivalam
TURAIYUR TALUK
To Turaiyur
MUSIRI TALUK
To Vriddha-chalam
To Padalur
PERAMBALUR TALUK
To Udayarpalaiyan
ARIYALUR TALUK
THANJAVUR DISTRICT
54.40
2.00
1.05
449.49
9.86
5.10
14.54
2.87
4.10
3.77
22.74
12.58
247.53
1.90
33.01
58.09
8.61
2.00
Lalgudi
To Tiruchirappalli
COLLEROON RIVER
: STUDY AREA
: DISTRICT BOUNDARY
: TALUK BOUNDARY
: VILLAGE BOUNDARY
: RAILWAY LINE
: ROADS
: RIVER
0 1 2 3 4 5 Miles
0 1 2 3 4 5 6 7 8 Kilometers
SCALE

Fig. 16

Poovalur (no. 64), 33 acres, both lying immediately to the south, and Siruganur (no. 91), four villages away in a northerly direction, 54 acres. Thus the external landholdings of the villagers of Peruvalanallur are very extensive and play an important role in the economy and external relations of the village.

Thus, Peruvalanallur villagers in 1980 owned 1,656.63 acres (wet: 861.81 acres; dry: 794.82 acres) of which 681.29 acres (wet: 582.14 acres; dry: 99.15 acres) or 41 percent were located within their own village, and the remaining 975.34 acres (wet: 279.67 acres; dry: 695.67 acres) or 59 percent were located in the other villages. Thus the villagers own land within the village, most of which has become wet land by expansion of irrigation over the last century, and even more land outside the village, most of which, however, is dry land.

Some important problems arise when someone holds land in a village other than his own. If the land is in the immediate neighboring villages, the owner regards the land as a part of his immediate farm unit, since the owner's land in the other village is within a manageable distance of his residence.[1] In a related manner, it should be noted that the villagers of the study area, which is generally known as a "wet village," tended to hold dry fields *(punjai)* in the neighboring "dry villages" and that, in contrast, the neighboring villagers tended to own wet fields *(nanjai)* in the study area. More specifically, although the Peruvalanallur villagers in 1980 owned large amounts of agricultural land in Kumulur (449.49 acres) and Edangimanglam (247.53 acres), immediately to the northwest, adjoining the dry lands of Peruvalanallur, these were overwhelmingly dry lands, 86 percent and 85 percent respectively. On the other hand, the villagers in Thachangurichi and Kumulur (both known as "dry villages") owned virtually no dry land in Peruvalanallur, but 62 acres and 36 acres of wet land respectively. As indicated previously, cropping patterns in the wet and dry lands are quite different. Thus, farmers who hold both types of land can be expected to cultivate more different types of crops and, by extension, to diversify their farming. The above feature of the complementary holdings of wet and dry lands among the villagers in and around their own villages influences the individual farmer's concept of how to utilize his lands and what type of farming to employ, which varies greatly depending upon the farmer's socio-economic situation; that is, some poor farmers want to sustain

---

1. It is unusual to cultivate fields whose location exceeds 4 kms (about 2.5 miles) from the owner's residence, although there are some exceptions.

at least their family consumption of food and other stuffs for their living from their own cultivation, and some rich farmers want to diversify their agricultural activities in order to have a greater production of commercial crops.

There is also another type of landholding whereby the Peruvalanallur villagers hold land in other distant villages. It is interesting to note that for this type of landholding the locations of the involved lands correspond mostly to the "mother villages" of the owner's family members in Peruvalanallur.[2] This is certainly related to the patterns of property inheritance and marriage in the rural communities concerned. Most villagers expressed the feeling that their sons and daughters have an equal right to share their parents' property, and this practice is regarded, at least on the surface, as the basic rule for property inheritance in the study area. There is also a customary rule that the mother's property such as land, called *sureedhana,* and ornaments brought to her husband as dowry are supposed to be given to her daughter(s). There are, however, some other rules applicable to the varied situations of parents and other members of the concerned families at the time of the property division. The fact that the practical ways of applying the rules vary greatly among the families has led to complicated patterns of property inheritance. This suggests that the rules themselves are not regarded as strict but loose.

In relation to the landholding under discussion, suffice it to say here that many women have registered their properties under their own names, most of which were inherited from their own parents. Thus, most of the female members' land in the remote areas was acquired through inter-village marriages.

Apart from the above case, migration with a family unit to the study area is also responsible for landholdings in the remote villages, since the migrants have usually retained their land in their "mother villages."

The scattered landholding in more remote villages resulting either from inheritance or migration generally does not form functional units of operation with the lands in the village and does not exhibit the same pattern of complementarity that characterizes the association of dry lands in adjacent villages with wet lands in Peruvalanallur.

---

2. Family members here include their ancestors, since occasionally land is left in the name of the deceased.

Sizes of Landholdings and Operational Farms

Because of the results of land transactions under certain types of land tenure, the size of each household's landholding is not necessarily the same as that of the operational farm. Since the problems involved in landownership and operation of a farm are interrelated, the data on both aspects were carefully checked and shown as accurately as possible. Figure 17 shows the percentages of households by size of landownership (A) and by operation (B) units and those of their corresponding ownership and management areas (C and D), with a distinction between wet and dry lands. More detailed data are shown in tables 10 and 11.

The size of landholdings of each household in Peruvalanallur in 1980 showed wide variation: from landless to 195.27 acres, with 1.90 acres being the average if the total area owned (1,656.63 acres) is divided by the total number of households (874). But 462 households in Peruvalanallur (53 percent) were in fact landless. If we take into account the landless and the very small landowners owning under one acre (165 households) or 1-2 acres (79 households), as many as 706 households (81 percent) together owned only 195 acres (12 percent) of the total area owned by the Peruvalanallur villagers (table 10). On the other hand, only 39 households (less than 5 percent of the households) owning 10-15 acres (21 households) and more than 15 acres (18 households) owned 890 acres, or about 54 percent of the total area. The distribution by size of holdings, both of number of families and of total area owned, is shown graphically in figure 17.

Figure 18 graphically depicts the uneven distribution of landownership and of land operation for all households and by individual castes. A straight line would indicate equal division of land among all families. All curves are notably concave indicating great variation in each caste in size of holdings by families. The first graph shows that nearly 50 percent of the households are without land and that the 20 percent of the largest landowners own about 90 percent of the land (i.e., the graphs start at about 50 percent cumulative households and cumulative 80 percent of households own about 10 percent of the land).

The contrasts among the three caste groups are very striking. The proportion of landless varies from less than 1 percent for the "forward" castes to about 50 percent for the "backward" castes and reaches 65 percent for the scheduled castes (*Harijans*). The cumulative percentages of households necessary to reach 10 percent of the owned land are also in contrast, about 50 percent for the

PERCENTAGE DISTRIBUTION OF HOUSEHOLDS AND AREAS BY SIZE OF LAND OWNERSHIP AND OPERATION IN PERUVALANALLUR VILLAGE (1980)

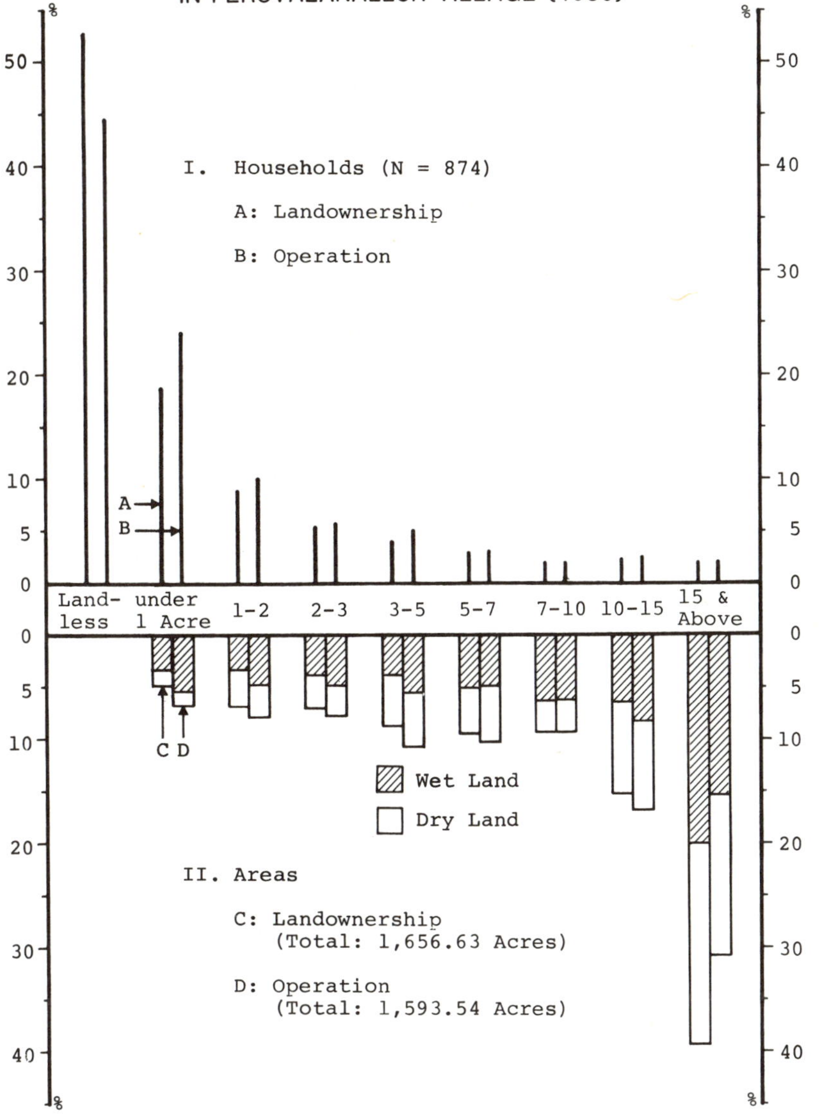

Fig. 17

TABLE 10

LANDOWNERSHIP: DISTRIBUTION OF HOUSEHOLDS AND THEIR CORRESPONDING AREAS BY CASTE AND SIZE OF LANDHOLDING IN PERUVALANALLUR VILLAGE, LALGUDI TALUK OF TIRUCHIRAPPALLI DISTRICT, TAMIL NADU, INDIA (1980)

Unit: in Acre

Names of Castes		Categories of Size of Landholding									Total
		1 Land-less	2 Under 1 Acre	3 1 - 2 Acres	4 2 - 3 Acres	5 3 - 5 Acres	6 5 - 7 Acres	7 7 - 10 Acres	8 10-15 Acres	9 15 Acres & Above	
I. Forward Castes											
1. Brahman	(H)	3	-	-	-	-	-	-	-	-	3
	(W)	-	-	-	-	-	-	-	-	-	-
	(D)	-	-	-	-	-	-	-	-	-	-
	(T)	-	-	-	-	-	-	-	-	-	-
2. Reddiar	(H)	4	2	8	8	9	7	15	9	17	79
	(W)	-	0.97	10.06	19.35	28.35	36.90	91.54	76.10	321.11	584.38
	(D)	-	0.07	1.36	0.74	7.85	6.33	38.36	37.51	297.78	390.00
	(T)	-	1.04	11.42	20.09	36.20	43.23	129.90	113.61	618.89	974.38
Sub-total (I)	(H)	7	2	8	8	9	7	15	9	17	82
	(W)	-	0.97	10.06	19.35	28.35	36.90	91.54	76.10	321.11	584.38
	(D)	-	0.07	1.36	0.74	7.85	6.33	38.36	37.51	297.78	390.00
	(T)	-	1.04	11.42	20.09	36.20	43.23	129.90	113.61	618.89	974.38
II. Backward (Middle Class) Castes											
3. Udaiyar	(H)	42	22	23	12	9	6	1	2	1	118
	(W)	-	8.86	12.57	12.35	12.73	16.08	2.25	1.02	11.62	77.48
	(D)	-	2.72	19.57	16.87	23.33	18.42	7.26	19.51	5.76	113.44
	(T)	-	11.58	32.14	29.22	36.06	34.50	9.51	20.53	17.38	190.92
4. Gounder	(H)	21	16	6	5	11	9	1	4	-	73
	(W)	-	4.54	2.54	4.89	16.71	17.91	2.16	18.21	-	66.96
	(D)	-	4.22	6.61	7.38	25.76	32.74	4.89	29.52	-	111.12
	(T)	-	8.76	9.15	12.27	42.47	50.65	7.05	47.73	-	178.08
5. Muslim	(H)	24	10	5	2	5	1	1	3	-	51
	(W)	-	4.85	3.93	4.21	5.00	2.04	7.17	3.27	-	30.47
	(D)	-	1.92	3.17	0.42	16.69	3.09	1.83	32.85	-	59.97
	(T)	-	6.77	7.10	4.63	21.69	5.13	9.00	36.12	-	90.44
6. Nadar	(H)	14	1	-	-	-	-	-	-	-	15
	(W)	-	0.45	-	-	-	-	-	-	-	0.45
	(D)	-	-	-	-	-	-	-	-	-	-
	(T)	-	0.45	-	-	-	-	-	-	-	0.45
7. Achari	(H)	9	4	1	-	-	-	-	-	-	14
	(W)	-	0.71	-	-	-	-	-	-	-	0.71
	(D)	-	0.03	1.00	-	-	-	-	-	-	1.03
	(T)	-	0.74	1.00	-	-	-	-	-	-	1.74
8. Muthuraja	(H)	7	5	2	-	-	-	-	-	-	14
	(W)	-	2.93	1.58	-	-	-	-	-	-	4.51
	(D)	-	0.40	1.15	-	-	-	-	-	-	1.55
	(T)	-	3.33	2.73	-	-	-	-	-	-	6.06

Notes: 1. The "forward", "backward", and "scheduled" castes are official categories employed by the Government of Tamil Nadu.

2. (H), (W), (D), and (T) respectively indicate household, wet land, dry land, and total area of wet and dry lands.

TABLE 10 (continued)

Unit: in Acre

Names of Castes		1 Land-less	2 Under 1 Acre	3 1 - 2 Acres	4 2 - 3 Acres	5 3 - 5 Acres	6 5 - 7 Acres	7 7 - 10 Acres	8 10-15 Acres	9 15 Acres & Above	Total
		Categories of Size of Landholding									
9. Naidu	(H)	10	-	-	-	-	-	-	-	-	10
	(W)	-	-	-	-	-	-	-	-	-	-
	(D)	-	-	-	-	-	-	-	-	-	-
	(T)	-	-	-	-	-	-	-	-	-	-
10. Pillai	(H)	8	-	-	-	-	-	-	-	-	8
	(W)	-	-	-	-	-	-	-	-	-	-
	(D)	-	-	-	-	-	-	-	-	-	-
	(T)	-	-	-	-	-	-	-	-	-	-
11. Vannan	(H)	4	1	2	-	-	-	-	-	-	7
	(W)	-	0.55	1.88	-	-	-	-	-	-	2.43
	(D)	-	-	1.06	-	-	-	-	-	-	1.06
	(T)	-	0.55	2.94	-	-	-	-	-	-	3.49
12. Pariyari	(H)	3	3	1	-	-	-	-	-	-	7
	(W)	-	1.13	-	-	-	-	-	-	-	1.13
	(D)	-	0.06	1.08	-	-	-	-	-	-	1.14
	(T)	-	1.19	1.08	-	-	-	-	-	-	2.27
13. Chettiar	(H)	3	1	1	-	-	1	-	-	-	6
	(W)	-	0.70	1.75	-	-	2.75	-	-	-	5.20
	(D)	-	-	-	-	-	3.04	-	-	-	3.04
	(T)	-	0.70	1.75	-	-	5.79	-	-	-	8.24
14. Pandaram	(H)	3	2	-	-	-	-	-	1	-	6
	(W)	-	-	-	-	-	-	-	4.14	-	4.14
	(D)	-	1.52	-	-	-	-	-	10.65	-	12.17
	(T)	-	1.52	-	-	-	-	-	14.79	-	16.31
15. Mooppanar	(H)	3	-	1	-	-	-	-	-	-	4
	(W)	-	-	-	-	-	-	-	-	-	-
	(D)	-	-	1.00	-	-	-	-	-	-	1.00
	(T)	-	-	1.00	-	-	-	-	-	-	1.00
16. Vellalar	(H)	3	-	-	-	-	-	-	-	-	3
	(W)	-	-	-	-	-	-	-	-	-	-
	(D)	-	-	-	-	-	-	-	-	-	-
	(T)	-	-	-	-	-	-	-	-	-	-
17. Agampadiar	(H)	3	-	-	-	-	-	-	-	-	3
	(W)	-	-	-	-	-	-	-	-	-	-
	(D)	-	-	-	-	-	-	-	-	-	-
	(T)	-	-	-	-	-	-	-	-	-	-
18. Muthaliar	(H)	1	1	-	-	-	-	-	-	-	2
	(W)	-	-	-	-	-	-	-	-	-	-
	(D)	-	0.10	-	-	-	-	-	-	-	0.10
	(T)	-	0.10	-	-	-	-	-	-	-	0.10
19. Padiyachi	(H)	2	-	-	-	-	-	-	-	-	2
	(W)	-	-	-	-	-	-	-	-	-	-
	(D)	-	-	-	-	-	-	-	-	-	-
	(T)	-	-	-	-	-	-	-	-	-	-

TABLE 10 (continued)

Unit: in Acre

Names of Castes		Categories of Size of Landholding									Total
		1 Land-less	2 Under 1 Acre	3 1 - 2 Acres	4 2 - 3 Acres	5 3 - 5 Acres	6 5 - 7 Acres	7 7 - 10 Acres	8 10-15 Acres	9 15 Acres & Above	
20. Christian	(H)	1	-	1	-	-	-	-	-	-	2
(Protestant)	(W)	-	-	-	-	-	-	-	-	-	-
	(D)	-	-	1.50	-	-	-	-	-	-	1.50
	(T)	-	-	1.50	-	-	-	-	-	-	1.50
21. Devar	(H)	1	-	-	-	-	-	-	-	-	1
	(W)	-	-	-	-	-	-	-	-	-	-
	(D)	-	-	-	-	-	-	-	-	-	-
	(T)	-	-	-	-	-	-	-	-	-	-
22. Konar	(H)	1	-	-	-	-	-	-	-	-	1
	(W)	-	-	-	-	-	-	-	-	-	-
	(D)	-	-	-	-	-	-	-	-	-	-
	(T)	-	-	-	-	-	-	-	-	-	-
23. Jangam	(H)	1	-	-	-	-	-	-	-	-	1
	(W)	-	-	-	-	-	-	-	-	-	-
	(D)	-	-	-	-	-	-	-	-	-	-
	(T)	-	-	-	-	-	-	-	-	-	-
24. Nayar	(H)	1	-	-	-	-	-	-	-	-	1
	(W)	-	-	-	-	-	-	-	-	-	-
	(D)	-	-	-	-	-	-	-	-	-	-
	(T)	-	-	-	-	-	-	-	-	-	-
Sub-total (II)	(H)	165	66	43	19	25	17	3	10	1	349
	(W)	-	24.72	24.25	21.45	34.44	38.78	11.58	26.64	11.62	193.48
	(D)	-	10.97	36.14	24.67	65.78	57.29	13.98	92.53	5.76	307.12
	(T)	-	35.69	60.39	46.12	100.22	96.07	25.56	119.17	17.38	500.60
III. Scheduled Castes (Harijans)											
25. Pallan	(H)	180	75	23	17	2	3	-	2	-	302
	(W)	-	24.55	16.75	21.84	1.42	8.58	-	4.11	-	77.25
	(D)	-	10.42	18.12	18.30	5.76	9.07	-	16.97	-	78.64
	(T)	-	34.97	34.87	40.14	7.18	17.65	-	21.08	-	155.89
26. Parayan	(H)	23	2	1	1	-	-	-	-	-	27
	(W)	-	0.33	-	-	-	-	-	-	-	0.33
	(D)	-	0.75	1.05	2.00	-	-	-	-	-	3.80
	(T)	-	1.08	1.05	2.00	-	-	-	-	-	4.13
27. Catholic	(H)	24	4	-	-	-	-	-	-	-	28
Pallan	(W)	-	1.22	-	-	-	-	-	-	-	1.22
	(D)	-	-	-	-	-	-	-	-	-	-
	(T)	-	1.22	-	-	-	-	-	-	-	1.22
28. Catholic	(H)	41	15	4	3	-	-	-	-	-	63
Parayan	(W)	-	1.90	0.97	1.68	-	-	-	-	-	4.55
	(D)	-	5.77	4.17	5.32	-	-	-	-	-	15.26
	(T)	-	7.67	5.14	7.00	-	-	-	-	-	19.81

TABLE 10 (continued)

Unit: in Acre

Names of Castes		Categories of Size of Landholding									Total
		1 Land- less	2 Under 1 Acre	3 1 - 2 Acres	4 2 - 3 Acres	5 3 - 5 Acres	6 5 - 7 Acres	7 7 - 10 Acres	8 10-15 Acres	9 15 Acres & Above	
29. Harijan Pariyari	(H)	2	-	-	-	-	-	-	-	-	2
	(W)	-	-	-	-	-	-	-	-	-	-
	(D)	-	-	-	-	-	-	-	-	-	-
	(D)	-	-	-	-	-	-	-	-	-	-
30. Harijan Dobi	(H)	3	-	-	-	-	-	-	-	-	3
	(W)	-	-	-	-	-	-	-	-	-	-
	(D)	-	-	-	-	-	-	-	-	-	-
	(T)	-	-	-	-	-	-	-	-	-	-
31. Ottan	(H)	4	-	-	-	-	-	-	-	-	4
	(W)	-	-	-	-	-	-	-	-	-	-
	(D)	-	-	-	-	-	-	-	-	-	-
	(T)	-	-	-	-	-	-	-	-	-	-
32. Domban	(H)	8	1	-	-	-	-	-	-	-	9
	(W)	-	0.60	-	-	-	-	-	-	-	0.60
	(D)	-	-	-	-	-	-	-	-	-	-
	(T)	-	0.60	-	-	-	-	-	-	-	0.60
33. Sukkilian	(H)	5	-	-	-	-	-	-	-	-	5
	(W)	-	-	-	-	-	-	-	-	-	-
	(D)	-	-	-	-	-	-	-	-	-	-
	(T)	-	-	-	-	-	-	-	-	-	-
Sub-total (III)	(H)	290	97	28	21	2	3	-	2	-	443
	(W)	-	28.60	17.72	23.52	1.42	8.58	-	4.11	-	83.95
	(D)	-	16.94	23.34	25.62	5.76	9.07	-	16.97	-	97.70
	(T)	-	45.54	41.06	49.14	7.18	17.65	-	21.08	-	181.65
TOTAL (I - III)	(H)	462	165	79	48	36	27	18	21	18	874
	(W)	-	54.29	52.03	64.32	64.21	84.26	103.12	106.85	332.73	861.81
	(D)	-	27.98	60.84	51.03	79.39	72.69	52.34	147.01	303.54	794.82
	(T)	-	82.27	112.87	115.35	143.60	156.95	155.46	253.86	636.27	1,656.63

Sources: The data consist of the author's Family Census for the village, the Ten-One Chitta, and the Adangals for 1978-79 and 1979-80.

TABLE 11

OPERATIONAL LANDS: DISTRIBUTION OF HOUSEHOLDS AND THEIR CORRESPONDING AREAS BY CASTE AND SIZE OF LANDHOLDING IN PERUVALANALLUR VILLAGE, LALGUDI TALUK OF TIRUCHIRAPPALLI DISTRICT, TAMIL NADU, INDIA (1980)

Unit: in Acre

Names of Castes		Categories of size of Landholding: 1 Land-less	2 Under 1 Acre	3 1 - 2 Acres	4 2 - 3 Acres	5 3 - 5 Acres	6 5 - 7 Acres	7 7 - 10 Acres	8 10-15 Acres	9 15 Acres & Above	Total
I. Forward Castes											
1. Brahman	(H)	3	-	-	-	-	-	-	-	-	3
	(W)	-	-	-	-	-	-	-	-	-	-
	(D)	-	-	-	-	-	-	-	-	-	-
	(T)	-	-	-	-	-	-	-	-	-	-
2. Reddiar	(H)	10	2	3	10	7	7	13	11	16	79
	(W)	-	1.47	3.94	20.80	24.86	28.14	71.08	94.86	224.77	469.92
	(D)	-	-	-	2.65	1.68	12.82	37.60	43.94	229.29	327.98
	(T)	-	1.47	3.94	23.45	26.54	40.96	108.68	138.80	454.06	797.90
Sub-total (I)	(H)	13	2	3	10	7	7	13	11	16	82
	(W)	-	1.47	3.94	20.80	24.86	28.14	71.08	94.86	224.77	469.92
	(D)	-	-	-	2.65	1.68	12.82	37.60	43.94	229.29	327.98
	(T)	-	1.47	3.94	23.45	26.54	40.96	108.68	138.80	454.06	797.90
II. Backward (Middle Class) Castes											
3. Udaiyar	(H)	34	27	26	8	11	8	1	2	1	118
	(W)	-	10.27	17.37	7.86	16.70	20.15	3.71	4.37	11.62	92.05
	(D)	-	3.21	18.65	11.33	22.98	26.09	3.85	16.67	5.76	108.54
	(T)	-	13.48	36.02	19.19	39.68	46.24	7.56	21.04	17.38	200.59
4. Gounder	(H)	23	14	7	3	12	9	2	2	1	73
	(W)	-	4.14	3.78	2.69	21.81	22.46	8.56	12.85	7.19	83.48
	(D)	-	3.03	6.19	5.12	25.34	30.54	7.57	10.82	10.11	98.72
	(T)	-	7.17	9.97	7.81	47.15	53.00	16.13	23.67	17.30	182.20
5. Muslim	(H)	25	11	3	4	3	1	1	3	-	51
	(W)	-	4.50	1.96	7.56	0.36	2.04	7.17	3.27	-	26.86
	(D)	-	2.35	1.75	1.39	13.38	3.09	1.83	32.85	-	56.64
	(T)	-	6.85	3.71	8.95	13.74	5.13	9.00	36.12	-	83.50
6. Nadar	(H)	11	3	1	-	-	-	-	-	-	15
	(W)	-	1.79	1.00	-	-	-	-	-	-	2.79
	(D)	-	-	-	-	-	-	-	-	-	-
	(T)	-	1.79	1.00	-	-	-	-	-	-	2.79
7. Achari	(H)	10	4	-	-	-	-	-	-	-	14
	(W)	-	1.11	-	-	-	-	-	-	-	1.11
	(D)	-	0.03	-	-	-	-	-	-	-	0.03
	(T)	-	1.14	-	-	-	-	-	-	-	1.14
8. Muthuraja	(H)	5	3	3	2	-	-	1	-	-	14
	(W)	-	1.93	3.71	4.19	-	-	8.00	-	-	17.83
	(D)	-	0.10	1.45	-	-	-	-	-	-	1.55
	(T)	-	2.03	5.16	4.19	-	-	8.00	-	-	19.38

Notes: 1. The "forward", "backward", and "scheduled" castes are official categories employed by the Government of Tamil Nadu.

2. (H), (W), (D), and (T) respectively indicate household, wet land, dry land, and total area of wet and dry lands.

TABLE 11 (continued)

Unit: in Acre

Names of Castes		Categories of Size of Landholding									Total
		1 Land-less	2 Under 1 Acre	3 1 - 2 Acres	4 2 - 3 Acres	5 3 - 5 Acres	6 5 - 7 Acres	7 7 - 10 Acres	8 10-15 Acres	9 15 Acres & Above	
9. Naidu	(H)	9	1	-	-	-	-	-	-	-	10
	(W)	-	0.28	-	-	-	-	-	-	-	0.28
	(D)	-	-	-	-	-	-	-	-	-	-
	(T)	-	0.28	-	-	-	-	-	-	-	0.28
10. Pillai	(H)	8	-	-	-	-	-	-	-	-	8
	(W)	-	-	-	-	-	-	-	-	-	-
	(D)	-	-	-	-	-	-	-	-	-	-
	(T)	-	-	-	-	-	-	-	-	-	-
11. Vannan	(H)	5	1	1	-	-	-	-	-	-	7
	(W)	-	-	0.43	-	-	-	-	-	-	0.43
	(D)	-	0.42	0.64	-	-	-	-	-	-	1.06
	(T)	-	0.42	1.07	-	-	-	-	-	-	1.49
12. Pariyari	(T)	2	5	-	-	-	-	-	-	-	7
	(W)	-	2.03	-	-	-	-	-	-	-	2.03
	(D)	-	0.71	-	-	-	-	-	-	-	0.71
	(T)	-	2.74	-	-	-	-	-	-	-	2.74
13. Chettiar	(H)	3	1	1	-	-	1	-	-	-	6
	(W)	-	0.70	1.75	-	-	2.75	-	-	-	5.20
	(D)	-	-	-	-	-	3.04	-	-	-	3.04
	(T)	-	0.70	1.75	-	-	5.79	-	-	-	8.24
14. Pandaram	(H)	1	2	2	-	-	-	-	1	-	6
	(W)	-	1.13	1.50	-	-	-	-	4.14	-	6.77
	(D)	-	-	1.02	-	-	-	-	10.65	-	11.67
	(T)	-	1.13	2.52	-	-	-	-	14.79	-	18.44
15. Mooppanar	(H)	3	1	-	-	-	-	-	-	-	4
	(W)	-	0.22	-	-	-	-	-	-	-	0.22
	(D)	-	-	-	-	-	-	-	-	-	-
	(T)	-	0.22	-	-	-	-	-	-	-	0.22
16. Vellalar	(H)	3	-	-	-	-	-	-	-	-	3
	(W)	-	-	-	-	-	-	-	-	-	-
	(D)	-	-	-	-	-	-	-	-	-	-
	(T)	-	-	-	-	-	-	-	-	-	-
17. Agampadiar	(H)	2	1	-	-	-	-	-	-	-	3
	(W)	-	0.39	-	-	-	-	-	-	-	0.39
	(D)	-	-	-	-	-	-	-	-	-	-
	(T)	-	0.39	-	-	-	-	-	-	-	0.39
18. Muthaliar	(H)	1	1	-	-	-	-	-	-	-	2
	(W)	-	-	-	-	-	-	-	-	-	-
	(D)	-	0.10	-	-	-	-	-	-	-	0.10
	(T)	-	0.10	-	-	-	-	-	-	-	0.10
19. Padiyachi	(H)	2	-	-	-	-	-	-	-	-	2
	(W)	-	-	-	-	-	-	-	-	-	-
	(D)	-	-	-	-	-	-	-	-	-	-
	(T)	-	-	-	-	-	-	-	-	-	-

TABLE 11 (continued)

Unit: in Acre

Names of Castes		Cagegories of Size of Landholding									Total
		1 Land-less	2 Under 1 Acre	3 1 - 2 Acres	4 2 - 3 Acres	5 3 - 5 Acres	6 5 - 7 Acres	7 7 - 10 Acres	8 10-15 Acres	9 15 Acres & Above	
20. Christian	(H)	2	-	-	-	-	-	-	-	-	2
(Protestant)	(W)	-	-	-	-	-	-	-	-	-	-
	(D)	-	-	-	-	-	-	-	-	-	-
	(T)	-	-	-	-	-	-	-	-	-	-
21. Devar	(H)	1	-	-	-	-	-	-	-	-	1
	(W)	-	-	-	-	-	-	-	-	-	-
	(D)	-	-	-	-	-	-	-	-	-	-
	(T)	-	-	-	-	-	-	-	-	-	-
22. Konar	(H)	1	-	-	-	-	-	-	-	-	1
	(W)	-	-	-	-	-	-	-	-	-	-
	(D)	-	-	-	-	-	-	-	-	-	-
	(T)	-	-	-	-	-	-	-	-	-	-
23. Jangam	(H)	1	-	-	-	-	-	-	-	-	1
	(W)	-	-	-	-	-	-	-	-	-	-
	(D)	-	-	-	-	-	-	-	-	-	-
	(T)	-	-	-	-	-	-	-	-	-	-
24. Nayar	(H)	1	-	-	-	-	-	-	-	-	1
	(W)	-	-	-	-	-	-	-	-	-	-
	(D)	-	-	-	-	-	-	-	-	-	-
	(T)	-	-	-	-	-	-	-	-	-	-
Sub-total (II)	(H)	153	75	44	17	26	19	5	8	2	349
	(W)	-	28.49	31.50	22.30	38.87	47.40	27.44	24.63	18.81	239.44
	(D)	-	9.95	29.70	17.84	61.70	62.76	13.25	70.99	15.87	282.06
	(T)	-	38.44	61.20	40.14	100.57	110.16	40.69	95.62	34.68	521.50
III. Scheduled Castes (Harijans)											
25. Pallan	(H)	134	104	30	18	11	2	-	3	-	302
	(W)	-	43.92	30.14	27.28	23.89	3.46	-	13.96	-	142.65
	(D)	-	8.44	14.33	14.42	16.55	7.67	-	18.37	-	79.78
	(T)	-	52.36	44.47	41.70	40.44	11.13	-	32.33	-	222.43
26. Parayan	(H)	19	5	2	1	-	-	-	-	-	27
	(W)	-	2.26	1.25	-	-	-	-	-	-	3.51
	(D)	-	-	1.05	2.00	-	-	-	-	-	3.05
	(T)	-	2.26	2.30	2.00	-	-	-	-	-	6.56
27. Catholic	(H)	14	10	3	1	-	-	-	-	-	28
Pallan	(W)	-	5.83	4.08	2.00	-	-	-	-	-	11.91
	(D)	-	-	-	-	-	-	-	-	-	-
	(T)	-	5.83	4.08	2.00	-	-	-	-	-	11.91
28. Catholic	(H)	35	15	7	5	1	-	-	-	-	63
Parayan	(W)	-	3.10	6.20	3.97	-	-	-	-	-	13.27
	(D)	-	4.23	3.82	7.94	3.98	-	-	-	-	19.97
	(T)	-	7.33	10.02	11.91	3.98	-	-	-	-	33.24

TABLE 11 (continued)

Unit: in Acre

Names of Castes		Categories of Size of Landholding 1 Land-less	2 Under 1 Acre	3 1 - 2 Acres	4 2 - 3 Acres	5 3 - 5 Acres	6 5 - 7 Acres	7 7 - 10 Acres	8 10-15 Acres	9 15 Acres & Above	Total
29. Harijan	(H)	2	-	-	-	-	-	-	-	-	2
Pariyari	(W)	-	-	-	-	-	-	-	-	-	-
	(D)	-	-	-	-	-	-	-	-	-	-
	(T)	-	-	-	-	-	-	-	-	-	-
30. Harijan	(H)	3	-	-	-	-	-	-	-	-	3
Vannan	(W)	-	-	-	-	-	-	-	-	-	-
	(D)	-	-	-	-	-	-	-	-	-	-
	(T)	-	-	-	-	-	-	-	-	-	-
31. Ottan	(H)	4	-	-	-	-	-	-	-	-	4
	(W)	-	-	-	-	-	-	-	-	-	-
	(D)	-	-	-	-	-	-	-	-	-	-
	(T)	-	-	-	-	-	-	-	-	-	-
32. Domban	(H)	9	-	-	-	-	-	-	-	-	9
	(W)	-	-	-	-	-	-	-	-	-	-
	(D)	-	-	-	-	-	-	-	-	-	-
	(T)	-	-	-	-	-	-	-	-	-	-
33. Sukkilian	(H)	5	-	-	-	-	-	-	-	-	5
	(W)	-	-	-	-	-	-	-	-	-	-
	(D)	-	-	-	-	-	-	-	-	-	-
	(T)	-	-	-	-	-	-	-	-	-	-
Sub-total (III)	(H)	225	134	42	25	12	2	-	3	-	443
	(W)	-	55.11	41.67	33.25	23.89	3.46	-	13.96	-	171.34
	(D)	-	12.67	19.20	24.36	20.53	7.67	-	18.37	-	102.80
	(T)	-	67.78	60.87	57.61	44.42	11.13	-	32.33	-	274.14
TOTAL (I - III)	(H)	391	211	89	52	45	28	18	22	18	874
	(W)	-	85.07	77.11	76.35	87.62	79.00	98.52	133.45	243.58	880.70
	(D)	-	22.62	48.90	44.85	83.91	83.25	50.85	133.30	245.16	712.84
	(T)	-	107.69	126.01	121.20	171.53	162.25	149.37	266.75	488.74	1,593.54

Sources: The data consist of the author's Family Census for the village, the Ten-One Chitta, and the Adangals for 1978-79 and 1979-80.

DISTRIBUTION OF LANDS BY CASTE GROUPS IN PERUVALANALLUR VILLAGE OF LALGUDI TALUK, TIRUCHIRAPPALLI DISRICT, TAMIL NADU, INDIA (1979-1980)

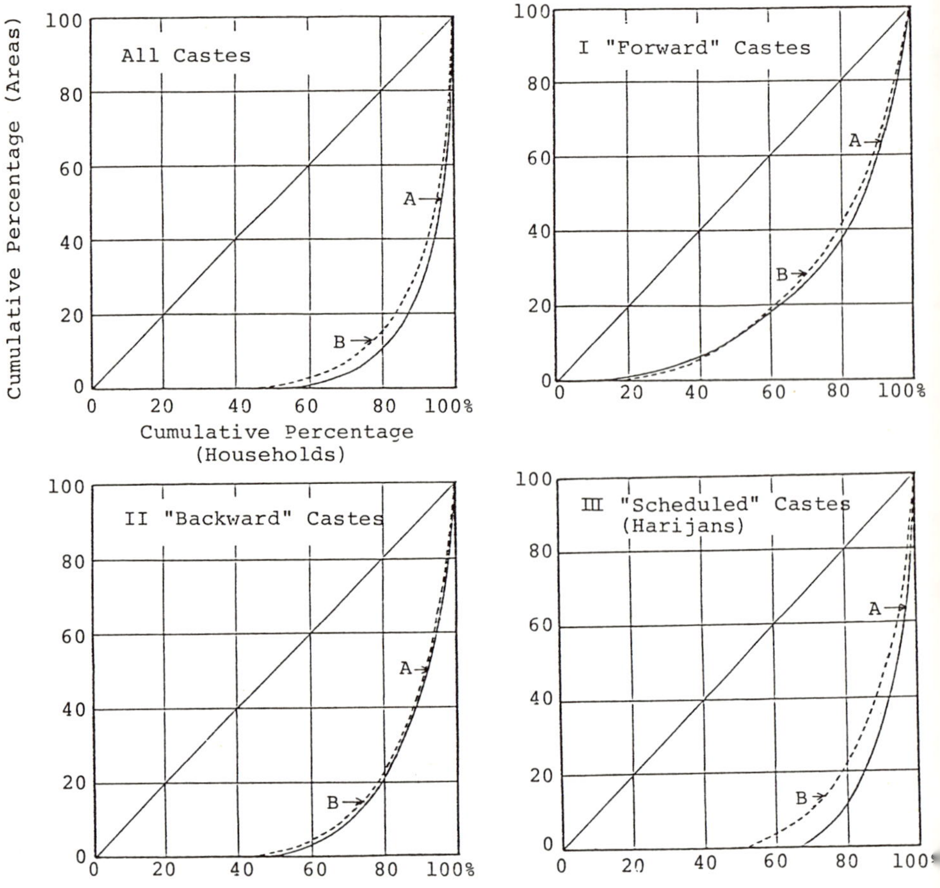

Notes: 1. The Curve A based on Lands under Ownership

2. The Curve B based on Lands under Operation

Fig. 18

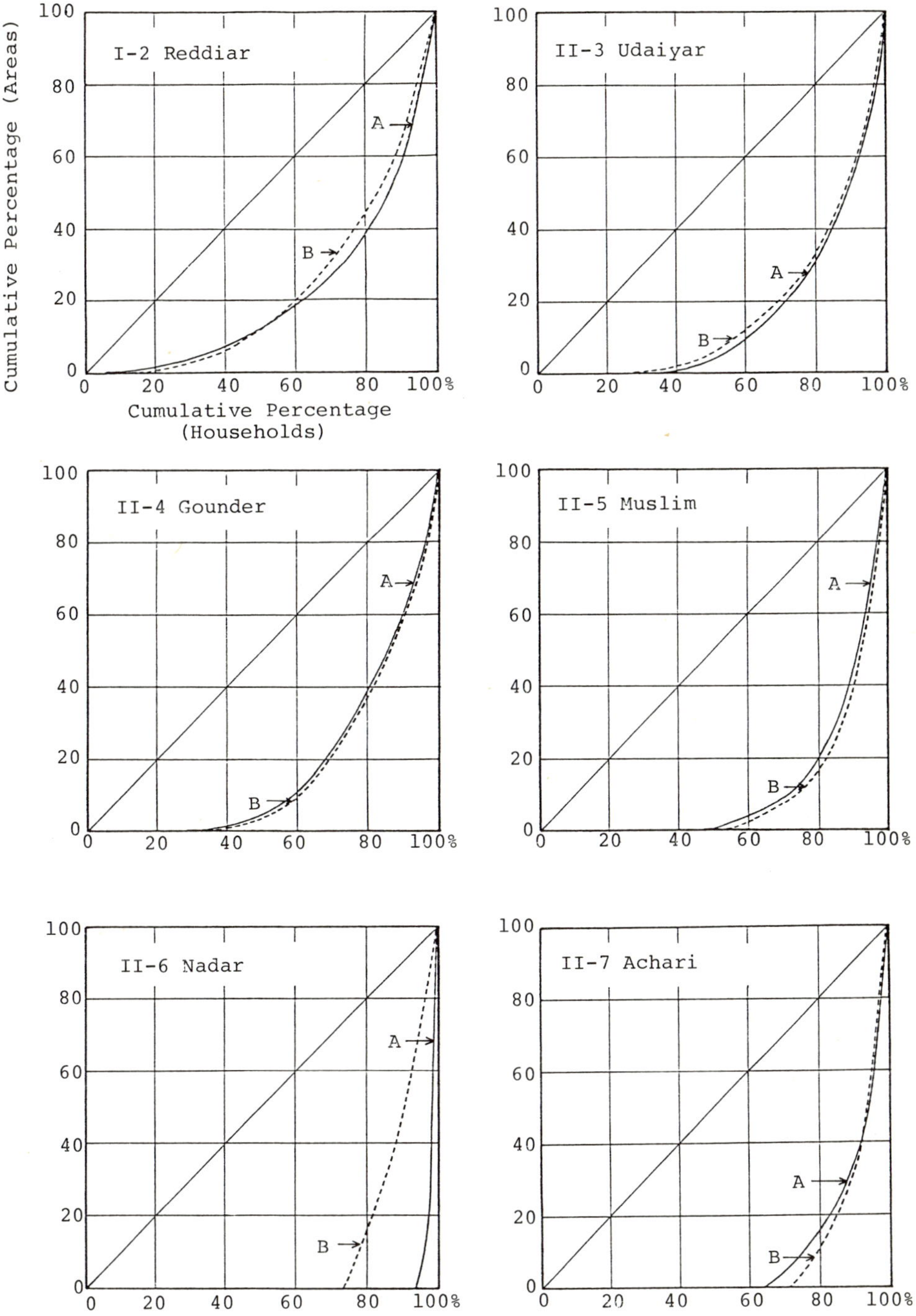
Cumulative Percentage (Areas)
Cumulative Percentage (Households)
I-2 Reddiar
II-3 Udaiyar
II-4 Gounder
II-5 Muslim
II-6 Nadar
II-7 Achari
A
B
0
20
40
60
80
100
100%

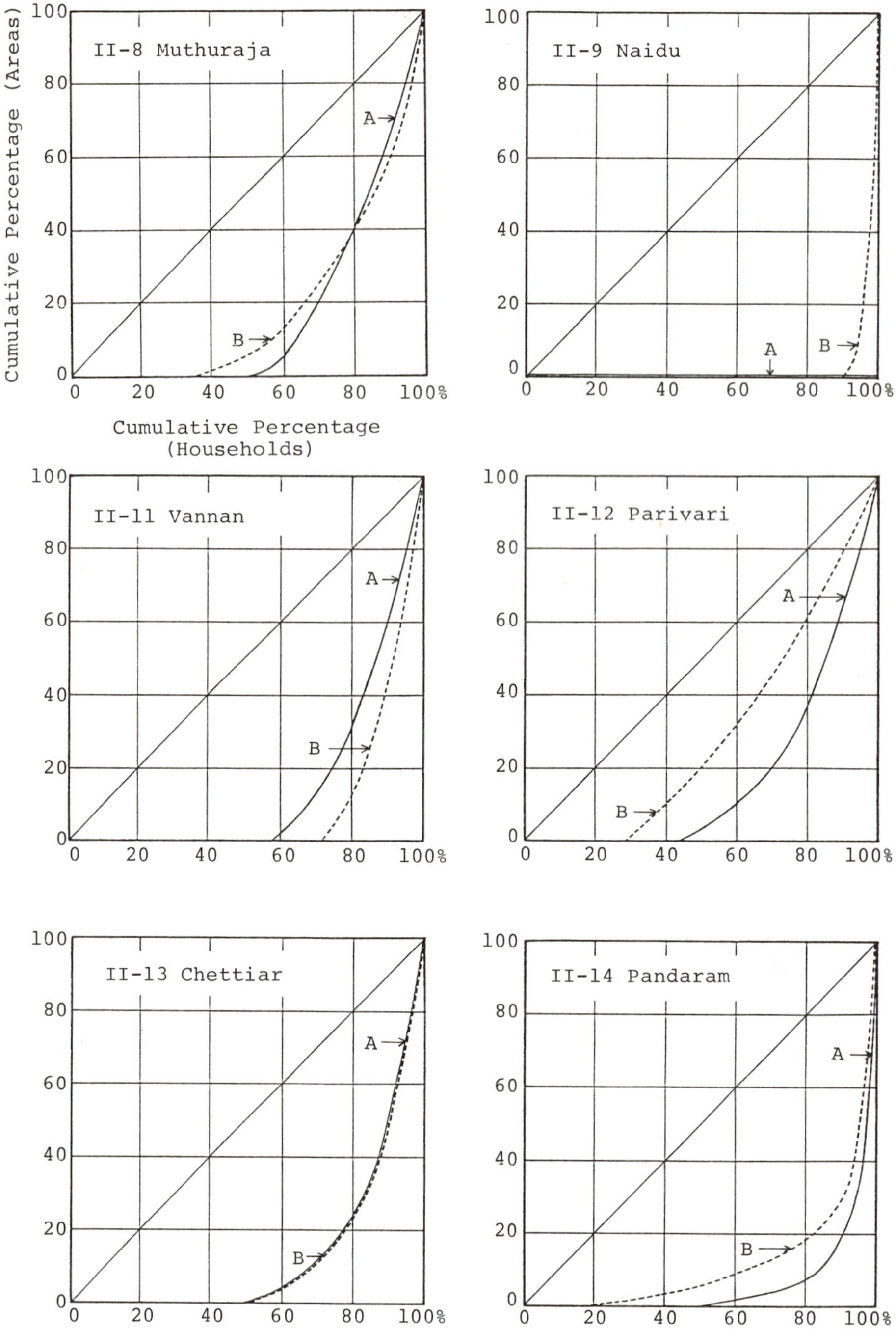
Cumulative Percentage (Areas)
II-8 Muthuraja
A
B
Cumulative Percentage (Households)
II-9 Naidu
A
B
II-11 Vannan
A
B
II-12 Parivari
A
B
II-13 Chettiar
A
B
II-14 Pandaram
A
B
0
20
40
60
80
100
100%

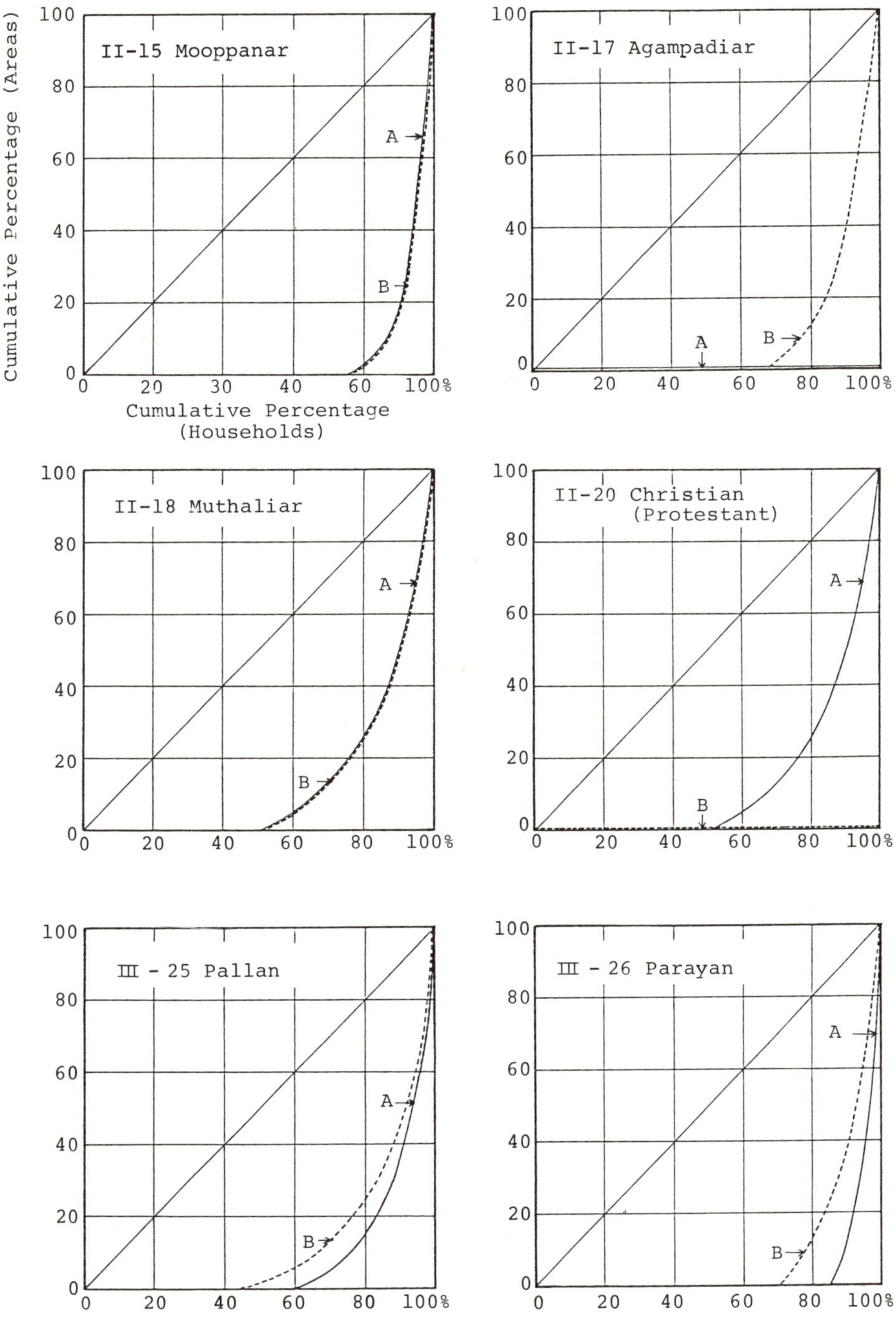
II-15 Mooppanar
Cumulative Percentage (Areas)
Cumulative Percentage (Households)
A
B
0
20
30
40
60
80
100
100%
II-17 Agampadiar
A
B
II-18 Muthaliar
A
B
II-20 Christian (Protestant)
A
B
III - 25 Pallan
A
B
III - 26 Parayan
A
B

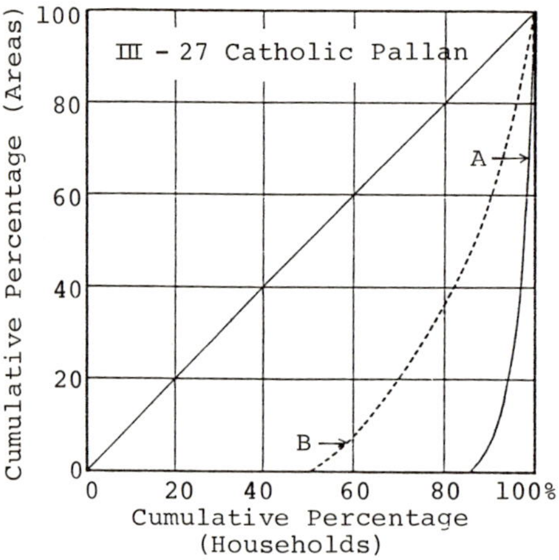

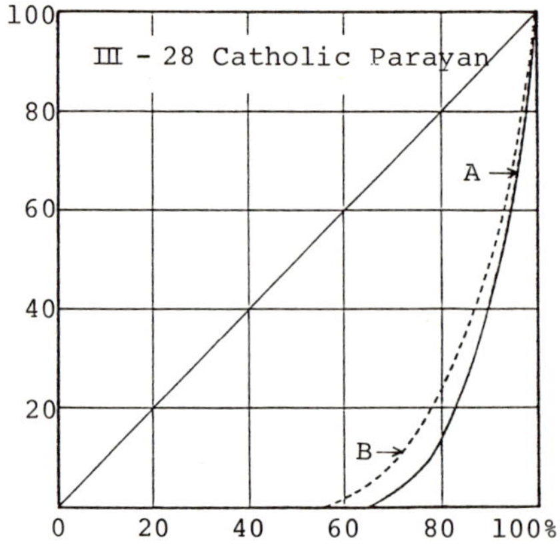

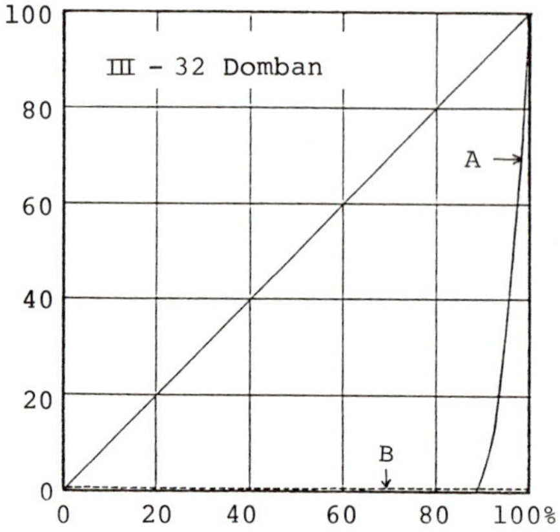

Notes: 1. The following 12 caste groups had virtually no land under their ownership and operation; (I-1) Brahman, (II-10) Pillai, (II-16) Vellalar, (II-19) Padiyachi, (II-21) Devar, (II-22) Konar, (II-23) Jamgam, (II-24) Nayar, (III-29) Harijan Pariyari, (III-30) Harijan Vannan, (III-31) Ottan, and (III-33) Sukkilian.

2. Each caste number in the above figures corresponds respectively to the number shown in Table 3

"forward" castes, about 70 percent for the "backward" castes, and about 80 percent for the scheduled castes. The curves of owned land and operated land are generally similar except among the scheduled castes which generally own little land but operate sizeable amounts of land owned by others.

Let us examine in more detail the size of the landholdings in relation to the different caste groups (table 10 and figs. 18 and 19). As indicate previously, the *Reddiars* (79 households) are regarded socio-economically as the leading caste group, and they owned 59 percent (974.8 acres) of the total area owned by the Peruvalanallur villagers, with 12.33 acres being the group's average. Furthermore the land owned by the *Reddiars* was generally of high quality, being predominantly wet land; indeed they owned 68 percent of the wet land owned by villagers (584 acres of a total 862) but only 49 percent of the dry land (390 acres of a total of 795). Stated another way, about 60 percent of the land owned by the *Reddiars* was wet land whereas about 60 percent of the land owned by all other castes was dry land. Among the *Reddiars*, 26 households held 10 acres or more, and together they owned 44 percent (732.50 acres) of the total area. Only 4 *Reddiar* households were landless, and only 10 *Reddiar* families owned 2 acres or less. After the *Reddiars*, the ranking of the landholding on a caste basis from the highest were the *Udaiyars*, 118 families holding 191 acres or nearly 12 percent of the total area, the *Gounders*, 73 families owning 178 acres or nearly 11 percent, the *Pallans*, one of the scheduled castes (*Harijans*), in which 302 households held 156 acres or 9 percent, and the Muslims, 51 families holding 90 acres or 5 percent, but their average holdings were much smaller, 1.62 acres, 2.15 acres, 0.52 acres, and 1.77 acres respectively.

Although the *Reddiars* owned most of the large landholdings, there were a few households of other castes which held 10 acres or more: 3 households in the *Udaiyars*, 4 in the *Gounders*, 3 in the Muslims, 2 in the *Pallans*, and 1 in the *Pandarams*. Out of the 33 caste groups in Peruvalanallur, 14 caste groups held virtually no agricultural land at all.

The operational lands of the village (table 11) are slightly less than the lands owned by villagers (table 10), 1,594 acres compared with 1,657 acres. The villagers operated 881 acres of wet land within the village owned by outsiders. They operated 713 acres of dry lands compared with 795 acres owned, suggesting that some of the scattered dry landholdings outside the village were not operated by the villagers. The distribution of operated land by caste was

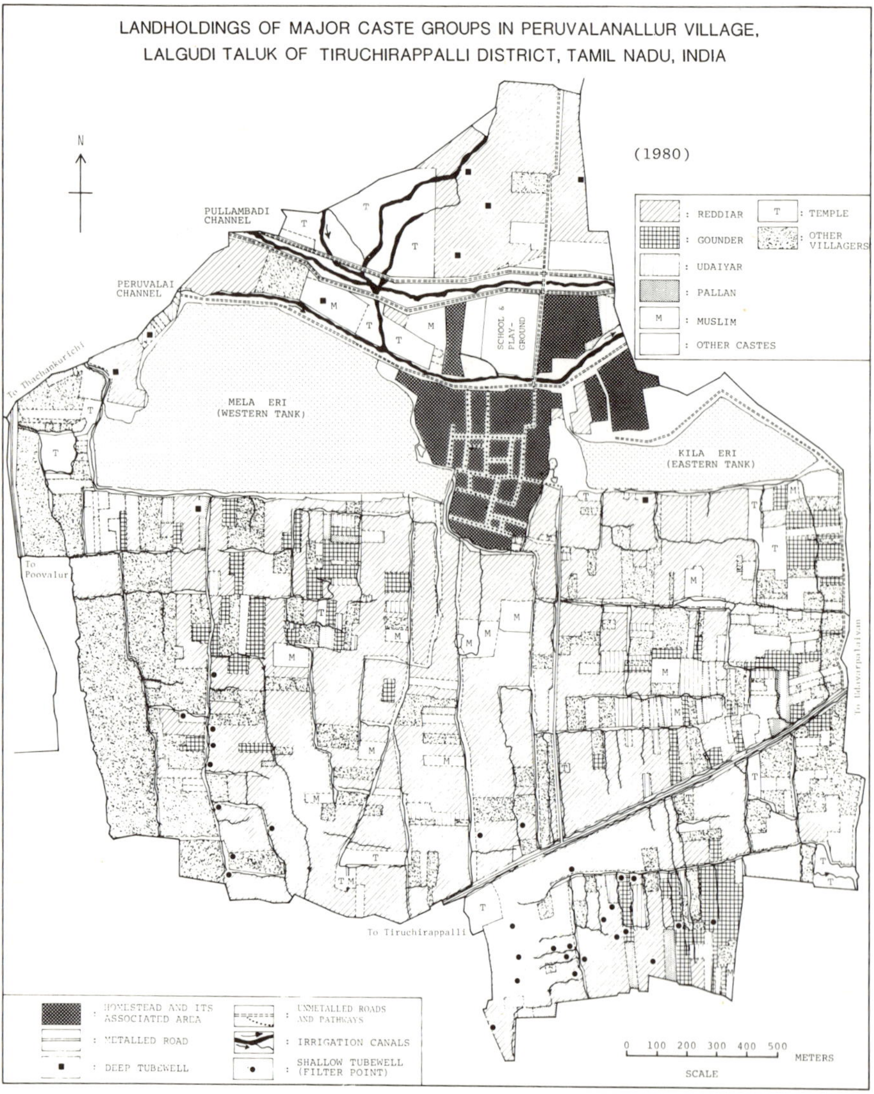
LANDHOLDINGS OF MAJOR CASTE GROUPS IN PERUVALANALLUR VILLAGE,
LALGUDI TALUK OF TIRUCHIRAPPALLI DISTRICT, TAMIL NADU, INDIA
(1980)
N
: REDDIAR
: GOUNDER
: UDAIYAR
: PALLAN
M : MUSLIM
: OTHER CASTES
T : TEMPLE
: OTHER VILLAGERS
PULLAMBADI CHANNEL
PERUVALAI CHANNEL
To Thachankurichi
MELA ERI (WESTERN TANK)
KILA ERI (EASTERN TANK)
SCHOOL & PLAY-GROUND
To Poovalur
To Udayarpalaiyam
To Tiruchirappalli
: HOMESTEAD AND ITS ASSOCIATED AREA
: METALLED ROAD
: DEEP TUBEWELL
: UNMETALLED ROADS AND PATHWAYS
: IRRIGATION CANALS
: SHALLOW TUBEWELL (FILTER POINT)
0 100 200 300 400 500 METERS
SCALE

Fig. 19

similar to that of owned land, but the role of the *Reddiars* was less dominant. They operated only 50 percent of the land in comparison with owning 59 percent of the land. Of particular importance were 114 acres of wet land owned but not operated by the *Reddiars* and 62 acres of dry land. The *Udaiyar, Gounder,* and *Pallan* households all operated substantially more wet land than they owned. The *Udaiyar, Gounder,* and Muslim households all operated somewhat less dry land than they owned. The sharpest contrast is for *Pallan (Harijan)* households that owned only 77 acres of wet lands but operated 143 acres.

As shown on figure 19 the dominant *Reddiar* caste owns a substantial part of the centrally located lands both of the large block of double-cropped wet land south of the residential area and the western and eastern tanks and of the dry lands immediately north of the residential areas. Three other leading castes *(Gounder, Udaiyar,* and Muslim) have a similar distribution but with scattered parcels only. The *Pallan (Harijan)* caste has land predominantly in the eastern strip of the wet lands only recently and partly transformed from single to double cropping. The holdings of other villagers lie in the western strip of wet lands still mainly in single cropping. The temple lands lie mainly just northeast of the western tank, partly double-cropped wet land partly dry land not agriculturally utilized in 1978-79 or 1979-80.

Transfers of landownership in the village during 1967-68 and 1979-80 consists of two types: inheritances and sales (fig. 20). Inherited areas in 184 cases transferred 192 acres, of which only 21 acres were dry land. The 197 sales transferred 173 acres, of which less than 11 were dry land.

With respect to inheritances, the *Reddiar* caste dominated with 115 of the inheritances or 62 percent, but with 148 acres or 77 percent of the total.

With respect to sales the *Reddiar* were again the dominant group accounting for 101 of 197 sales, and 104 acres of a total of 173 acres (60 percent of the area sold) and virtually all of the modest amount of sales of dry land (about 10 acres). The *Reddiar* purchased only 32 acres of the 162 acres of wet land sold, or just under 20 percent. There were no purchases by *Reddiar* of dry lands. About 40 percent of all land sold represented a net transfer from *Reddiar* to other castes, amounting to 72 acres (104 sold, 32 purchased). The principal net purchasers were other villages with a net gain of 28 acres (45 sold and 73 purchased). The Muslim *(Labai)* group had a net gain of 15 acres (17 bought and 2 sold), the *Udaiyar* a net gain of 9 acres (16 bought and 7 sold), and *H. Pallan,* 8 acres (13 bought and 5 sold). These are net transfers, but there is some tendency for members of a caste to sell to other members of the same caste; thus more than 50 percent of all sales by *Udaiyar, Gounder,* and *H. Pallan* were to other members of the same caste. Some 65 percent of the sales of other villagers were to other villagers.

TRANSFERS OF LANDOWNERSHIP BY CASTE DURING 1967-68 AND 1979-80 IN THE PERUVALANALLUR VILLAGE OF LALGUDI TALUK, TIRUCHIRAPPALLI DISTRICT, TAMIL NADU, INDIA

Total Transfers:	365.00 Acres
A. Inheritances:	192.33 Acres
B. Sales:	172.67 Acres

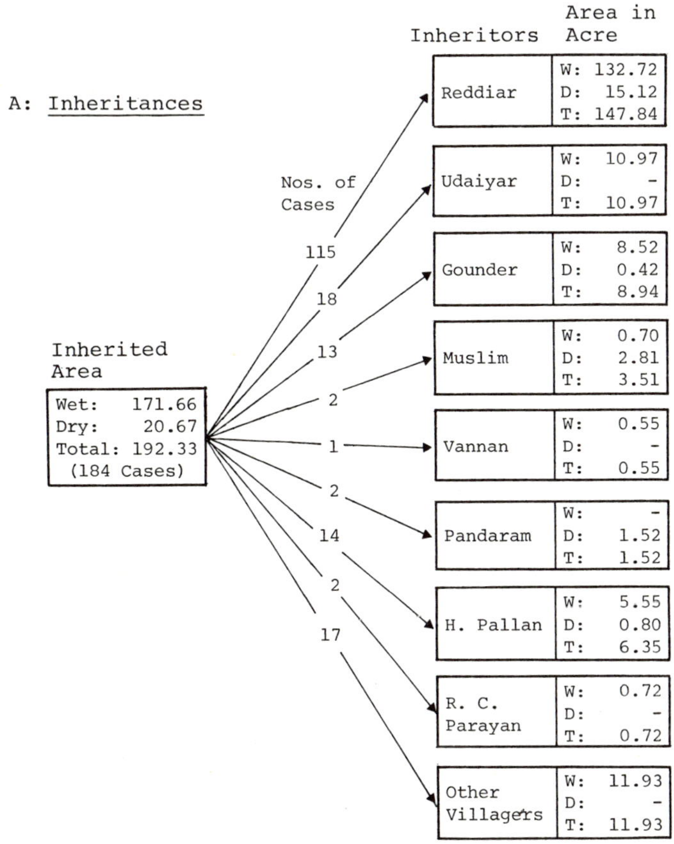

Fig. 20

TRANSFERS OF LANDOWNERSHIP

B: Sales

Previous Owners: Reddiar — Wet: 94.17, Dry: 9.83, Total 104.00 (101 Cases)

Nos. of Cases	Present Owners	Area in Acre W:	D:	T:
28	Reddiar	22.15	1.88	24.03
11	Udaiyar	9.92	-	9.92
11	Gounder	7.38	-	7.38
11	Muslim (Labai)	8.94	7.55	16.49
3	Muthuraja	1.56	-	1.56
6	H.Pallan	4.32	-	4.32
1	R.C. Parayar	0.20	-	0.20
30	Other Villagers	39.70	0.40	40.10

Previous Owners: Udaiyar — Wet: 6.96, Dry: -, Total: 6.96 (19 Cases)

Nos. of Cases	Present Owners	Area in Acre W:	D:	T:
2	Reddiar	0.83	-	0.83
9	Udaiyar	3.52	-	3.52
5	Gounder	1.33	-	1.33
2	H.Pallan	1.14	-	1.14
1	Other Villagers	0.14	-	0.14

Previous Owners: Gounder — Wet: 7.06, Dry: -, Total: 7.06 (13 Cases)

Nos. of Cases	Present Owners	Area in Acre W:	D:	T:
1	Reddiar	1.17	-	1.17
8	Gounder	3.89	-	3.89
1	H.Pallan	0.46	-	0.46
3	Other Villagers	0.86	-	0.86

Previous Owners: Muslim (Labai) — Wet: 1.78, Dry: -, Total: 1.78 (3 Cases)

Nos. of Cases	Present Owners	Area in Acre W:	D:	T:
2	H.Pallan	1.54	-	1.54
1	Other Villagers	0.92	-	0.92

TRANSFER OF LANDOWNERSHIP

B: <u>Sales</u>

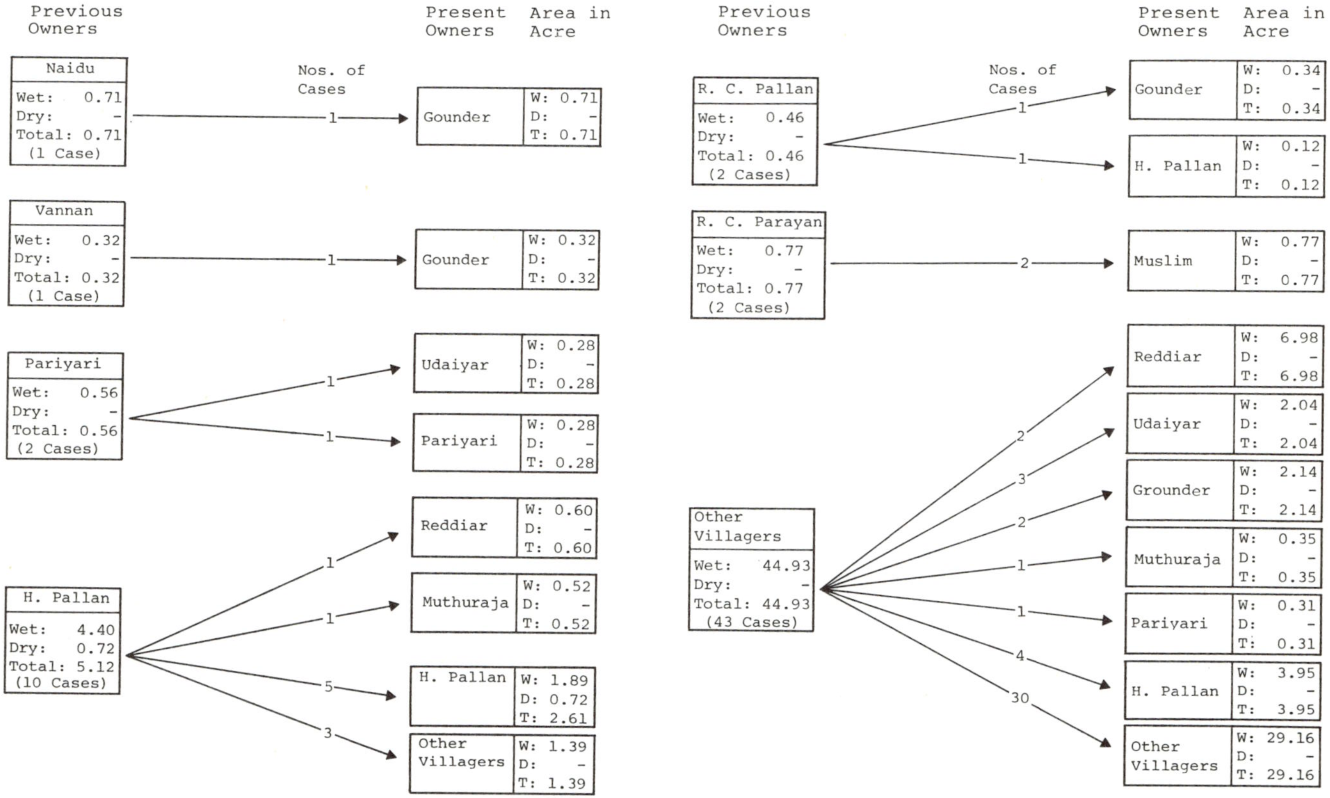

CHAPTER V
OCCUPATIONAL SPECIALIZATION AND LABOR ORGANIZATION

Occupational Specialization

Working Population

Working population means here those persons engaged in certain varieties of labor in exchange for tangible rewards. It is more useful to point out the borderline cases than to elaborate upon this minimal definition for our present purpose. The seasonal variations of land utilization often create predictable short-term and long-term underemployment or unemployment among those who engage in agriculture. Employed or self-employed agricultural workers therefore include those who are potentially capable of, and/or are willing to engage in, agriculture if given opportunities regularly. The same applies to the non-agricultural workers.

Although there are rough age limits to the working population, the two categories of minors and the aged have been included. At around the age of 10-13, boys in some families are given regular minor chores such as cattle tending, and they assist in plowing, planting, and harvesting. They are not counted as working members as such. The exceptions are those who live as resident servants attached to families other than their own and receive wages and/or rewards in kind (such as paddy) and those who are regularly entrusted with the handling of a plow or other important agricultural implements without adult supervision. Those who attend school are not included in the working population regardless of age.

Old men generally do not engage regularly in heavy physical labor; but those who are active in supervising and coordinating labor and other resource allocations, especially in agriculture, are counted as members of the working population. The exact distinction is no doubt subject to the writer's personal judgment with regard to working and non-working youngsters, but he believes that the exclusion of all the youngsters and old men from the working population statistically creates more distortion than their partial inclusion would.

Women are counted as part of the working population as long as they are hired for tangible rewards. Otherwise their labor is considered part of the housekeeping tasks for women as wives, mothers, and daughters. Thus, these women are not included in the working population. For instance, their participation in agricultural activities is often limited to those chores which are also allocated to non-working youngsters. The kinds of tasks done by women are admittedly an important part of the division of labor for managing agricultural and other occupational activities, but this attribute is regarded as not directly pertinent to the topic of this study.

Occupational Specialization

The classification of occupations for working members in the study area provides the basic data for the following discussion. However, it is an extremely difficult task to arrange all the working members into limited categories in the agricultural and non-agricultural sectors. This is primarily because the sectors are not rigidly separated but inter-penetrate each other: the same individual may be partly engaged in both sectors and may have to make decisions which, to varying degrees, affect his commitment in each of them.[1] An example would be the man who is primarily a businessman but participates in farming as a supervisor, since he owns many operational lands; or the man who is a factory worker but cultivates some land. It has also been observed that the same individual can be partly engaged in some of the different types of work within the agricultural sector. Thus, the classification of occupations is possible only when we disregard some of the less important overlapping parts of the individual activities within a sector or between sectors.

Table 12 shows the number of male and female working members listed by caste groups in the seven categories (A-G) of the agricultural sector. For convenience sake, this table also includes the total figures of the number of male and female working members under the non-agricultural sector and the percentage of the agricultural workers from the total working members within each caste group.

---

1. S.B. Yamey, "The Study of Peasant Economic Systems: Some Concluding Comments and Questions" in Raymond Firth and S.B. Yamey, *Capital, Saving and Credit in Peasant Society* (London: George Allen & Unwin Ltd., 1963), p. 378.

TABLE 12

OCCUPATIONAL CLASSIFICATION IN PERUVALANALLUR VILLAGE, LALGUDI TALUK OF TIRUCHIRAPPALLI DISTRICT, TAMIL NADU, INDIA (1980)

(1980)

Names of Castes	I. Agricultural Occupations (A: Supervisors B: Owner-cultivators C: Owner-cultivators with Wage-laborers D: Landless Wage-laborers E: Pannaiyals F: Pannaikarans G: Shepherds)														1 Sub-total		2 Total	II. Non-agricultural Occupations 3 Sub-total		4 Total	5 Grand Total	6 2/5x100
	A		B		C		D		E		F		G									
	M	F	M	F	M	F	M	F	M	F	M	F	M	F	M	F	M+F	M	F	M+F	2+4	(%)
I. Forward Castes																						
1. Brahman	-	-	-	-	-	-	-	-	-	-	-	-	-	-	-	-	-	7	4	11	11	-
2. Reddiar	55	3	-	-	-	-	-	-	-	-	-	-	-	-	55	3	58	47	1	48	106	54.72
Sub-total	55	3	-	-	-	-	-	-	-	-	-	-	-	-	55	3	58	54	5	59	117	49.57
II. Backward (Middle Class) Castes																						
3. Udaiyar	7	1	23	6	40	65	15	24	5	-	9	-	-	-	99	96	195	20	-	20	215	90.70
4. Gounder	3	-	25	2	11	-	17	13	4	-	10	-	3	-	73	15	88	20	-	20	108	81.48
5. Muslim	6	2	1	-	-	2	1	4	-	-	-	-	-	-	8	8	16	58	-	58	74	21.62
6. Nadar	-	-	-	-	4	4	9	8	-	-	-	-	-	-	13	12	25	2	-	2	27	92.59
7. Achari	-	-	-	-	-	-	1	3	-	-	-	-	-	-	1	3	4	17	1	18	22	18.18
8. Muthraja	-	-	9	-	2	-	3	-	-	-	-	-	-	-	14	-	14	7	-	7	21	66.67
9. Naidu	-	-	-	-	-	-	9	6	-	-	1	-	-	-	10	6	16	2	-	2	18	88.89
10. Pillai	-	-	-	-	-	-	2	2	-	-	-	-	-	-	2	2	4	7	-	7	11	36.36
11. Vannan	-	-	-	-	-	-	-	-	-	-	-	-	-	-	-	-	-	8	10	18	18	-
12. Pariyari	-	-	-	-	-	4	-	1	-	-	-	-	-	-	-	5	5	7	-	7	12	41.67
13. Chettiar	-	-	-	-	-	-	-	1	-	-	-	-	-	-	-	1	1	8	-	8	9	11.11
14. Pandaram	1	-	-	1	5	7	1	1	-	-	-	-	-	-	7	9	16	2	-	2	18	88.89
15. Mooppanar	-	-	-	-	-	-	1	5	1	-	-	-	-	-	2	5	7	-	-	-	7	100.00
16. Vellalar	-	-	-	-	-	-	-	-	-	-	-	-	-	-	-	-	-	3	-	3	3	-
17. Agampadiar	-	-	-	-	-	-	-	2	-	-	-	-	-	-	-	2	2	3	-	3	5	40.00
18. Muthaliar	-	-	-	-	-	-	-	1	-	-	-	-	-	-	-	1	1	2	-	2	3	33.33

TABLE 12 (continued)

Names of Castes	I. Agricultural Occupations A: Supervisors B: Owner-cultivators C: Owner-cultivators with Wage-laborers D: Landless Wage-laborers E: Pannaiyals F: Pannaikarans G: Shepherds														1 Sub-total		2 Total	II. Non-agricultural Occupations			5 Grand Total	6 2/5x100
	A		B		C		D		E		F		G					3 Sub-total		4 Total		
	M	F	M	F	M	F	M	F	M	F	M	F	M	F	M	F	M+F	M	F	M+F	2+4	(%)
19. Padaiyachi	-	-	-	-	-	-	2	-	-	-	-	-	-	-	2	-	2	-	-	-	2	100.00
20. Christian	-	-	-	-	-	-	-	-	-	-	-	-	-	-	-	-	-	2	1	3	3	-
21. Devar	-	-	-	-	-	-	-	-	-	-	-	-	-	-	-	-	-	1	-	1	1	-
22. Konar	-	-	-	-	-	-	-	-	-	-	-	-	-	-	-	-	-	1	-	1	1	-
23. Jangam	-	-	-	-	-	-	-	-	-	-	-	-	-	-	-	-	-	1	-	1	1	-
24. Nayar	-	-	-	-	-	-	-	-	-	-	-	-	-	-	-	-	-	1	-	1	1	-
Sub-total	17	3	58	9	62	82	61	71	10	-	20	-	3	-	231	165	396	172	12	184	580	68.28
III. Scheduled Castes (Harijans)																						
25. Pallan	-	-	5	6	138	177	152	166	9	-	7	-	-	-	311	349	660	55	4	59	719	91.79
26. Parayan	-	-	-	-	1	10	12	26	-	-	-	-	-	-	13	36	49	13	-	13	62	79.03
27. (Catholic) Pallan	-	-	-	-	14	18	16	12	-	-	-	-	-	-	30	30	60	7	1	8	68	88.24
28. (Catholic) Parayan	-	-	-	-	28	35	35	41	1	-	-	-	-	-	64	76	140	31	1	32	172	81.40
29. (Harijan) Pariyari	-	-	-	-	-	-	1	3	-	-	-	-	-	-	1	3	4	3	-	3	7	57.14
30. (Harijan) Vannan	-	-	-	-	-	-	-	3	-	-	-	-	-	-	-	3	3	3	-	3	6	50.00
31. Ottan	-	-	-	-	-	-	2	2	-	-	-	-	-	-	2	2	4	1	-	1	5	80.00
32. Domban	-	-	-	-	-	-	13	9	-	-	-	-	-	-	13	9	22	3	-	3	25	88.00
33. Sukkilian	-	-	-	-	-	-	5	5	-	-	-	-	-	-	5	5	10	-	-	-	10	100.00
Sub-total	-	-	5	6	181	240	236	267	10	-	7	-	-	-	439	513	952	116	6	122	1,074	88.64
Total	72	6	63	15	243	322	297	338	20	-	27	-	3	0	725	681	1,406	342	23	365	1,771	79.39
Grand Total		78		78		565		635		20		27		3		1,406	1,406		365	365	1,771	79.39

The categories of agricultural workers in table 12 and the corresponding numbers of working members are as follows:

I-A.	Supervisors	78 persons
I-B.	Owner-cultivators	78
I-C.	Owner-culltivators with Wage-Laborers .	565
I-D.	Landless Wage-laborers	635
I-E.	*Pannaiyals*	20
I-F.	*Pannaikarans*	27
I-G.	Shepherds	3
	Total Agricultural Workers	1,406

Category I-A. Supervisors.
Working members who are classified in this category seldom participate in the physical work of the agricultural activities but are mostly in charge of such managerial tasks as hiring agricultural laborers and allocating tasks to working family members and hired hands. Such members in the study area are mostly senior household heads and/or those with relatively large landholdings in their households, though there are a few exceptions.

Category I-B. Owner-cultivators. Although this type of agriculturalist exhibits seasonal variations in labor utilization, he does not as a rule seek outside employment during the agricultural slack seasons. He is sometimes found in the same families with the category I-C type who partly or fully hire out their labor as agricultural laborers. It can be generally assumed in such cases that the size of the family landholding is less than that which is manageable by the available family labor force.

Category I-C. Owner-cultivators and Category I-D, Landless Wage-laborers. Agricultural laborers usually belong to families without, or with only marginal, landholdings, or otherwise to families which have more of a labor force than necessary for the cultivation of their land. If we further classify the 565 working members under category I-C into two groups depending upon the size of the family landholdings, i.e., (1) I-C-1: two acres or more of land and (2) I-C-2: less than 2 acres of land, the working members in each group are 238 persons (98 males and 140 females) and 327 persons (145 males and 182 females) respectively. Between the two groups under category I-C, the latter members (I-C-2) in general need more work than the former (I-C-1). However, it is generally true that landless laborers (category I-D) are the most serious about seeking work opportunities; in fact, they work not only within the study area but also in distant villages.

Category I-E. Pannaiyals.

The *pannaiyal* is an agricultural laborer hired on an annual basis, who is regarded as the most intimate and reliable worker among the hired laborers because of his frequent associations with his employer and the rest of the family members. The *pannaiyal* has to show up every morning and evening to get his employer's instructions for the daily work. Thus, the *pannaiyals* are mostly hired from the same village or the neighboring villages within commuting distances of the employers. Only when they are hired from remote villages, do they live in the employer's compound. The *pannaiyals* work is almost like the resident laborer in other societies, from milking to managerial work in the agricultural activities. The payment system for the *pannaiyal* in the study area is a unique one. He gets 2 *kalams* (1 *kalam* is about 50 kgs) of unhusked paddy per month; however, he receives only 1 *kalam* at the end of each month and his employer keeps the other 1 *kalam* for each month and the total (12 *kalams* a year) is eventually given at one time to the *pannaiyal* at the end of the year. (This is a subtle way to control the *pannaiyal.)* Although the *pannaiyal* system has been commonly practiced in this part of Tamil Nadu, this payment practice, as stated above, seems to be of recent origin and is practiced only in this limited area.[2]

Category I-F. Pannaikarans.

The *pannaikaran* is a young servant (10-15 years of age), who is usually hired on an annual basis. Because of their age, the *pannaikarans* generally do not do more than domestic errands and minor agricultural tasks. They are mostly from poor landless families, which find some relief in their being fed by employers as well as in their small earnings. Their wages are determined according to the employee's age, capability, and personality (such as honesty and obedience).

Category I-G. Shepherd.

Three persons listed under this category belong to the *Gounder* caste whose occupation is shepherding. The *Gounders* usually keep sheep and goats to a lesser or greater extent, and their livelihood is

---

2. The payment system in some villages of the Tiruchy District is quite different from that in the study area. See Venkatesh B. Athreya et al., *A Study of Production Relations in Agriculture (Midterm Report)* (Tiruchy: Madras Institute of Development Studies and Lund University, December 1979), p. 31 (Mimeographed).

partly dependent upon the livestock which are sold at weekly markets or available meat shops within Peruvalanallur. It is interesting to note that mature sheep or goats which are kept as a group (usually 20-30 in number) in others' fields over night can yield one rupee per animal per month for their dung.

At this point, let us examine some important relationships between the distribution of the working population under the particular categories stated above and the different caste *(jati)* groups. In this respect, we should consider the figures of each group in table 12 along with the basic statistics for each group shown in tables 3 and 11 in the previous chapters. Out of 1,771 persons of the total working members of the Peruvalanallur village in 1980, 1,406 persons or 79.4 percent were in the agricultural sector. If we include such members as the cartmen,[3] owners and/or co-owners of the "milk society,"[4] and those who are engaged in both the agricultural and non-agricultural sectors (explained earlier), the above percentage would easily exceed 80 percent. In any event, there is a great variation in the ratio of the agricultural members to the non-agricultural ones among the different caste groups.

We cannot deal with occupational specialization of each caste group based simply on the ratio of agricultural or non-agricultural members, because some caste groups have only small numbers of households from which to draw meaningful generalizations. However, as an example of such minor caste groups, it should be noted that some technical caste groups such as the *Achari* (blacksmith, carpenter, etc.) and service caste groups such as the *Vannan* (washerman) and *Pariyari* (barber) still engage in their traditional occupations, although some members of the former are factory workers. This kind of example can be observed in some other minor caste groups.

Among the sizeable caste groups, the Muslims showed an extremely low ratio in the agricultural sector (22 percent). This is closely related to their recent temporary migrations to the oil-producing Gulf countries for better job opportunities. This kind of movement has occurred among the Muslim communities on a national scale since the early 1970s. Out of the 74 Muslim working

---

3. The cartmen here are regarded as hired laborers, since they neither have a pair of bullocks nor an "improved cart" which is equipped with tires. The improved cart was introduced in recent years and is used mainly for business purposes. The owners of the cart usually hire the cartmen whose work is regarded to be equivalent to those of a truck driver.

4. The "milk society" is a privately organized body at the village level in which the organizer (owner has a contract with each of the milk cow owners (usually 5-10 members) to buy the fresh milk at a fixed rate. The cow owners have to bring their cows every morning and evening to the cow-sheds of the society owners or to any assigned place for milking.

members in all Peruvalanallur, 40 persons have left the village for this kind of work:[5] 31 persons to Kuwait, one person each to Dubai and Muscat (Oman), 2 persons to Malaysia, and 6 persons to Bombay. The overseas working members' salaries vary greatly from Rs. 1,500 for factory watchmen to Rs. 5,000 for truck drivers. These salaries are nearly 10 times the regular pay scale in the study area.

These Muslim overseas activities have had a great socio-economic impact not only on their own families but also on Peruvalanallur village as a whole and the wider locality as well. Since the workers have left their family members in the village, it is generally true that their wives have assumed the great burden of the family responsibility and of the farm management.[6] Under these circumstances, the concerned family members have had to face various kinds of difficulties and troubles of which the most unhappy case resulted in a family's break-up and divorce. The Muslim economic power, however, has been influential in many aspects of the rural communities in and around the study area. One of the most important aspects is that they have been interested in buying agricultural land; this seems to have been responsible for the higher price of land in recent years. The general acceptance of this Muslim "buying power" has not only raised land prices but has also been used as a justification for raising land values. Thus though the Muslims represented only 4 percent of the working population, they accounted for about 10 percent of all land purchases and about 20 percent of those purchases which represented a net transfer from one caste group to another.

Among the sizeable caste groups, after the Muslims, the *Reddiars* showed a lower ratio (55 percent) of agricultural working members, although their operational lands together accounted for 50 percent (797.90 acres) of the village total (1,656.63 acres). As a landowning caste group, the *Reddiars* have had better opportunities for securing occupations in the non-agricultural sector compared with other caste groups, since they are better acquainted with job markets and have their own economic power to establish self-employed businesses. However, this does not necessarily mean that the *Reddiars* can find jobs easily. In fact, there are several university and college graduates, including the author's assistants, who had been looking for permanent jobs for two years after their

---

5. During their term of employment, the married men have left their family members behind in the village.

6. It was interesting for the author to observe a Muslim woman who, with an umbrella in the fields, was supervising the hired women laborers in paddy transplanting. Later on, it was learned that her husband had been in Kuwait for the last two years.

graduation.

Regarding the *Reddiars* involvement in the agricultural sector, it should be mentioned that their working members (58 persons) fall exclusively under category I-A, supervisors. What are the reasons for this kind of *Reddiar* participation in agricultural activities? It is obviously related to the fact that the *Reddiars* in general hold many acres of land under their individual ownership. It is also a fact that there are many marginal landowners within the *Reddiars*. Part of an answer lies in the social context. Many *Reddiars* do not go into their fields, and most of them do not even touch the soil. The author found only four persons who do some light physical work in the fields and lead cattle. This kind of *Reddiar* behavior appears to maintain their self-appraised role as members of the leading caste in Peruvalanallur. The above behavior, however, can not be applied to the same caste group in other villages, i.e., Reddimangudi (no. 35 in table 1 and on fig. 4), Neykulam (no. 89), and Garudamangalam (no. 104) in the dry zone, where the *Reddiars* do not regard "touching the soil and leading cattle and other animals" as a dishonor. In fact, even in Peruvalanallur, the above behavior is surely of recent origin, because there is evidence that some of the grandfathers of the present leaders of the *Reddiar* community had migrated to this village as *pannaiyals* (agricultural laborers on an annual basis).

Unlike the above Muslim and *Reddiar* caste groups, most other sizeable caste groups under both "backward" and scheduled castes showed very high ratios (above 80 percent) in the agricultural working population (table 12). However, they are highly varied in their distribution under the stated categories of the agricultural sector. This is primarily reflected by the size of landownership and/or operational areas of each caste group (cf. tables 10 and 11).

There are great differences in the participation of men and women in the various occupational groups. Thus supervisors and owner-cultivators are overwhelmingly men (72 men supervisors compared with 6 women, 63 men owner-cultivators compared with 15 women). Among non-agricultural occupations men outnumber women by 342 to 23. The minor agricultural occupations of *pannaiyal* (agricultural laborers hired on an annual basis), *pannaikarans* (young servants), and shepherds consist solely of males. Among the agricultural wage laborers (categories I-C and I-D of table 12) of the village as a whole, however, the number of women workers (660) substantially exceeds the number of male workers (540). Women workers in this category represent 37 percent of the working population of the village. Insofar as women are counted as part of the working population (excluding housekeeping), 94 percent are engaged as agricultural wage laborers. It should be noted that the wages of women agricultural workers are considerably lower than that of men, a phenomenon which will be discussed in the following section.

Labor Organization

Regardless of the size of the landholdings of any given farmhouse, there are usually managing members within their families responsible for the agricultural operations. However, it appears that the major physical work for agricultural activities on the individual farms is largely dependent upon the hired laborers, even with the marginal farmhouses which have such work members in their own families. The reason is that most of the manual work for agriculture is generally conducted intensively and collectively in a limited period on the individual farms by employing laborers. The employers or their agents usually do not have contact with the individual laborers, but with their leader who procures the necessary numbers of laborers and transmits the conditions for the work to the rest of the laborers. Thus, the hired laborers temporarily form a "work party" organized by a leader through whom the daily wages are distributed to each member of the group.

Under this practice, if a marginal farmer works some days on his own land together with his hired laborers, he (the employer) is regarded as a member of the "work party" and gets his formal share of the payment for the day. The employer in this example might be hired as an agricultural laborer along with the other laborers for some other day on other farms.

There are two common wage systems for the agricultural day laborers: (1) the ordinary fixed wage system, and (2) the contract labor system.[7]

It should be noted that some types of work are assigned only to male laborers whereas some other types only to female laborers (table 13).

The agricultural laborers in and around the study area worked 5 hours a day which consisted of a morning time (9:00-12:00) and an evening time (4:00-6:00).

Male laborers were paid Rs. 5 (Rs. 3 for the morning time and Rs. 2 for the evening) per day, while females are paid only Rs. 2.25 (Rs. 1.50 for the morning and Rs. 0.75 for the evening). This wage

---

7. This system is locally called *"kothu"* in some other *taluk*s of Tiruchy District. See Athreya et al., op. cit., p. 34.

TABLE 13

SPECIFIC WORKS IN AGRICULTURAL ACTIVITIES ASSIGNED TO MALE AND FEMALE LABORERS

Agricultural Work	Male	Female
Sign (X): Applicable; Sign (0): Not-applicable		
I. General Work		
1. Ploughing and leveling	X	0
2. Re-digging ditches for field canals, and watching irrigation and drainage	X	0
3. Transplanting for both paddy and sugar-cane	0	X
4. Weeding	X	X
5. Spraying pesticides and chemical fertilizers	X	0
6. Manuring	X	0
II. Harvesting the Paddy		
1. Cutting	X	X
2. Bundling	X	X
3. Carrying to the threshing grounds	X	X
4. Threshing	X	X
5. Helping the thresher	0	X
6. Winnowing	0	X
III. Harvesting the Sugarcane		
1. Cutting	X	0
2. Bundling	X	X
3. Carrying to the roads	X	X

system applies to all work done except paddy and sugar cane harvests. Only for paddy harvesting is an equal wage in kind, i.e., paddy, given to both male and female laborers.

The contract labor system is always applied to paddy and sugar cane harvests in the study area. As is the nature of the contract system, the number of laborers in a party may vary from 4 or 5 to 25 or 30 depending upon the work assigned by the employer to be completed that day.

It is interesting to note that there are certain combinations of male and female laborers. The number of male and female workers participating in the paddy harvest is equal, while for the sugar cane harvest, the number of males and females is usually at the rate of 1:2.

For paddy harvest in this system, a party of the hired laborers is assigned to complete a series of work including cutting, bundling, carrying to the threshing grounds, threshing, winnowing, straw-piling, and weighing the grain. For this work the party of laborers gets a fixed ratio of the total harvest completed by the party during the day. The ratio itself, however, varies depending upon the varying distances between the fields and the threshing grounds; that is, the ratios between the landowner and the laborers' party are 10:1 for a distance under 150 meters, 8:1 for a distance from 150 to 300 meters, and 6:1 for a distance beyond 300 meters.

For the sugar cane harvest in the contract system, the party of the laborers is usually assigned to complete a series of work including cutting, bundling, carrying to the roads, and loading on tractors or carts. The party of laborers is paid at a fixed rate per metric ton for the total harvest completed by the party during the day. As in the case of paddy harvest, the distance between the fields and the nearest roads is an important factor in determining the rate per ton for the party; it is paid at the rate of Rs. 15 per ton for a distance under 150 meters, and for further distances Rs. 5 per ton is added for every 100 meters.

Individual male and female laborers in a party under the contract system usually enjoy a higher wage of Rs. 7-9 and Rs. 4-6 per day respectively. This might be related to the result of the collective bargaining power of the laborers in the system in that there are cultivators' time schedules which have to be met. One of them is with sugar cane where the cultivator must reserve a time at the sugar cane factory. If the cultivator does not bring his sugar cane by the appointed time, the factory threatens not to accept it. There are also limitations on the harvest period. This is especially true for paddy. However, it should be noted that the individual laborers in this system not only achieve more work per hour but also extend their working time by 1-3 hours more per day than those working in the ordinary fixed wage system, because they receive a percentage of the produce.

CHAPTER VI
LAND TENURE AND ITS IMPLICATIONS

Land Tenure and Tenant Regulations

There are three types of land tenure currently available in Tamil Nadu, namely, (1) *varam* (share-cropping tenure), (2) *kuttagai* (fixed rent tenure), and (3) *otti* (usufructuary mortgage tenure), although the last might not be properly called tenure.[1] Each type of tenure has its own long history, but may not have begun as the same system which is presently practiced. When we focus on the current land tenures, it should be noted that the above three types cannot be observed to an equal extent in all the villages, *taluks*, or districts of Tamil Nadu. Rather, it is more common that, among the three types of tenure only one type or two are dominantly practiced in a given village or area. In fact, in Peruvalanallur, only *kuttagai* and *otti* are observed and dominantly practiced. Before proceeding to a detailed discussion of various aspects of land tenure in the studied village, it might be useful to outline some characteristics of each type of tenure, and also some of the important tenancy acts and rules which are relevant to the current changes in land tenure in rural communities in Tamil Nadu.

Types of Land Tenure

Varam is share-cropping tenancy. It dates back at least to the Chola empire of the ninth to thirteenth centuries and probably long before.[2] In the present *varam* system, the amount of rent is determined by the total gross produce in the involved land and share-ratios between the landlord and his tenant. The share-ratios of return between the landlord and his tenant vary greatly, depending not only upon the different provisions for irrigation water for the involved lands, but also upon the cost-bearing

---

1. Joan P. Mencher, *Agricultural and Social Structure in Tamil Nadu, Past Origins, Present Tranformations and Future Prospects* (New Delhi: Allied Publishers Private Ltd., 1978), p. 96.

2. Kathleen Gough, *Rural Society in Southeast India* (Cambridge: Cambridge University Press, 1981), p. 47.

conditions between the two parties for kinds of agricultural work, fertilizers and pesticides, seeds, operational work on and maintenance of the irrigation facilities such as *aetram eiravai* (slope-lifting by people), *kavalai* (skin-bag-lifting by a pair of bullocks), low-lift pumps with diesel engine power, and deep tubewells with electric power.

Kuttagai seems to be the most common type of tenure in Tiruchy District. It has a long history, as the term *kuttagai* was said to be introduced into Thanjavur by the Vijayanagar rulers sometime between the mid-fourteenth and mid-seventeenth centuries.[3] Under the current *kuttagai* system the tenant is supposed to pay a fixed rent in cash or kind to his landlord, the amount is settled before cultivation (the rent, however, is paid after the harvest), and the tenant has to bear all the cultivation expenses. In the wet villages where different types of paddy *(kuruvai, thaladi,* and *samba),* sugar cane, and banana are major crops, the rent is paid in kind for paddy cultivation and in cash for sugar cane and banana cultivation. In the dry villages, cultivable lands are mostly *punjais* (nonirrigated lands), which are used for several types of pulses and millets, groundnuts, chillies, vegetables, etc. The rent for such lands (*punjais*) is usually paid in cash regardless of the kinds of crops cultivated. Even in the dry villages there are some pockets of *nanjais* (wet lands), which are used mostly for paddy cultivation. The rent for such lands is usually paid in kind. However, the landowners in the dry villages usually do not like leasing their wet lands under the *kuttagai* tenure because these lands are the most productive within the dry environ. They prefer to cultivate these limited wet lands by themselves, if possible. It should be noted that, in many cases in the *kuttagai* transaction, there is a difference in varying degrees between the fixed amount and the actual amount of payment of rent, since most of the tenants "bargain" with the landowners after the harvest.

The fixed rent per unit of area varies greatly not only within a village but also in the different villages or regions, mainly because of the various qualities of the *kuttagai* lands which provide different land productivities. In the wet villages like Peruvalanallur, the amount of rent is determined largely by the criterion for the "single-cropping" and "double-cropping" lands. In 1979-80 in Peruvalanallur, the rent per acre for paddy cultivation in the "single-cropping" wet lands was 7.5 bags - 12.0 bags of unhusked rice, or 40-60 percent of the gross produce (one bag is

---

3. Ibid., pp. 47-48.

about 60 kg), and in "double-cropping" wet lands, 10.5 - 15.0 bags, or 24-34 percent of the crop. In the case of sugar cane, the rent per acre was Rs. 600 - Rs. 1,000, or 20-35 percent of the "net income."

The *otti* type of tenure also has a long history, since the term *otti* appeared in the temple inscriptions in the early fifteenth century.[4] Under the current *otti* system, a tenant gets the right of cultivation of land involved by depositing a certain amount of cash in advance with his landlord. The period of the *otti* contract is usually for three years. The full right of the land involved is returned to the landowner on the repayment of the deposit without any interest to the tenant. Thus, we can regard the tenant's (creditor's) yearly enjoyment of cultivating the *otti* land as an "annual interest" on the cash deposit to the landowner (debtor).

Tenancy Regulations

Since the 1950s, the legislature of Tamil Nadu has passed several tenancy acts and rules, and their amendments, which are aimed mainly at protecting the "cultivating tenant."[5] Of these acts, the following three acts are worth mentioning: (1) the Tamil Nadu (Madras) Cultivating Tenants Protection Act, 1955; (2) the Tamil Nadu (Madras) Cultivating Tenants (Payment of Fair Rent) Act, 1956; and (3) the Tamil Nadu Agricultural Lands Records of Tenancy Right Act, 1969 and their amendments.

The 1955 Act protects the tenants from all kinds of arbitrary evictions, retrospectively from December 1, 1953, but enables the landowners, in deserving cases, to repossess land for personal cultivation. However, very few tenants and landowners have resorted to the courts;[6] the majority of cases have been settled out of court one way or another. The results were described by Mencher as follows:[7]

> By far the majority of tenants in the village we know simply gave up their tenancies. Some tried to get tenancies from other landlords. Some were even shifted by the same landlord to another plot. Protests were rare.

---

4. Noboru Karashim and Y. Subbarayalu, "Varangai/Indangai, Kaniyalar, and Irajagarattar: Social Conflict in Tamilnadu in the 15th Century," *Socio-cultural Change in Villages in Tiruchirappalli District, Tamilnadu, India, Part 1 (Pre-modern Period),* ed. Noboru Karashima (Tokyo: Institute for Study of Languages and Cultures of Asia and Africa, 1981), p. 144.

5. According to the tenancy acts, the "cultivating tenant" means a person who contributes his own physical labor or that of any member (heir) of his family in the cultivation of any land belonging to another under a tenancy agreement, expressed or implied. The acts distinguish the cultivating tenant from a mere intermediary or his heir and is aimed at protecting the former only.

6. K.S. Sonachalam, *Land Reforms in Tamil Nadu* (New Delhi: Oxford and IBH Pub. Co., 1970), pp. 32-62.

7. Mencher, op. cit., p. 111.

The 1956 Act attempts to abolish usury amd rack-renting. There are three aspects which fix the fair rent: (1) the classification of land into three categories, where irrigation (its nature and intensity) is the most important criterion; (2) the determination of the normal gross produce for each category; and (3) the fixing of a percentage of the gross produce payable as fair rent for each category of land in which the percentage is correlated in each case inversely to the irrigation intensity.[8] This act fixes fair rent as follows: (1) in the case of wet land, 40 percent of the normal gross produce or its value in money; (2) in the case of wet land where the irrigation is supplemented by lifting water, 35 percent of the total gross produce or its value in money; and (3) in the case of any other class of land, 33.3 percent of the total gross produce or its value in money. In the case of lands in items (1) and (2), in which water is lifted by pumpsets installed at the cost of the landowners, the fair rent can be increased to 40 percent. With respect to the above, a great many villagers and scholars express their opinions that the actual rate of rent of the total produce paid by the tenants is much higher than the fixed rate of rent supported by the act. This may be generally true, but the measurement of the rate of rent should be carefully looked at. In the case of paddy cultivation, for example, the villagers are likely to report the "total gross produce" after subtracting some of the paddy which is paid to the hired laborers as their wages.[9]

The 1969 Act was passed with a view toward regulating the working of the other acts which were intended to protect the tenants. The act provides for drawing up a land record of rights, which includes such particulars as the survey (or subdivision) number, area, names and addresses of owners and tenants, etc. Special officers are appointed to investigate, verify, and record these tenancy agreements. The implementation of the act was in two stages: for the districts of Tanjavur, Tiruchy, and Madurai, it went into effect on September 8, 1971. Thus, the first Gazettes of the Tenancy Registers under the 1969 Act were published in August 1972; although further applications for the Tenancy Registers are still open to the tenants concerned.

---

8. G. Venkataramani, *Land Reform in Tamil Nadu* (Madras: Madras Institute of Development Studies, 1973), p. 29.

9. In the case of the *kuruvai* paddy cultivation, the harvesting work is usually done by a group of hired laborers for a certain unit area on a contractual basis, and the group gets a fixed share of the total harvest (real gross produce). See chapter V.

The Gazettes have included only some of the actual cases of the land tenure transactions. The reasons for this have been mainly either the tenants' innocence or hesitancy to file the applications in the Tenancy Registers, because of their passive attitude toward the 1969 Act. The tenants' attitude itself seems to be related to the speculation or perception that, even if the application were filed, there would be an unfair rejection of the entry in the Tenancy Registers instigated by the landowners. It is believed, however, that, once a tenancy agreement is listed in the Tenancy Registers, the landowner practically has to give up the right of cultivation of the land almost "permanently," since the act protects not only the present tenant but also his heirs. Therefore, the landowners have tried to resist, by all means, entering their land tenure in the Tenancy Registers.

Most of the above passive attitude by the tenants is rooted in their rather recent basic belief that the land is already theirs (and their heirs') to cultivate permanently. Thus, they see no point in going through the complicated rigors of registering the lands they cultivate. This attitude applies only to the *kuttagai* tenancy in the studied area of the Tiruchy District.

Although quite a few scholars say that the above tenancy acts are not effective, this writer feels that they have neglected to look at the actual influence of these acts. For example, it has been observed in the study area that most of the *kuttagai* tenants and landowners know that the tenants have a "permanent" right of cultivation of the lands; the tenants now often feel no need to register those lands; and the landowners, who are short of working members in their own families, are reluctant to buy any more land--they prefer to invest their capital in other ways. Thus, it is obvious that the spirit of these acts, if not their actual letter, is increasingly influential, especially in the wet villages.

Varam in Lalgudi Taluk

Although *varam* is not practiced at all in Peruvalanallur,[10] it was found in some nearby dry villages in Lalgudi Taluk, including Mahizambadi (no. 33 in table 1 and on figure 4), Reddimangudi (no. 35), Neykulam (no. 89), and Siruganur (no. 91). Some of the share-ratios between the landlord and his tenants and their respective cost-bearing conditions are exemplified as follows:[11]

---

10. *Varam* tenancy is also not practiced at all in Garudamangalam and Alunthalaipur (both under no. 104).

11. Tsukasa Mizushima and Tsuyoshi Nara, "Social Change in a Dry Village in South India," *Studies in Socio-Cultural Change in Rural Villages in Tiruchirapalli District in Tamilnadu, India, no. 4* (Tokyo: Institute for Study of Languages and

1. One-third of the gross produce to the landlord and two-thirds to the tenant. In this type, locally called *nel-varam* or *nil-varam,* the tenant alone takes care of the cropping and all the cultivating expenses. The lands are usually *punjai*s (dry lands) without any irrigation facilities and rain-fed only.[12]
2. One-half to the landlord and one-half to the tenant. In this type, locally called *alipathi-varam* (half-each-*varam*), the landlord usually pays only for manure and fertilizers, and the tenant bears all the rest of the cultivating expense. As in the first case, the lands are usually *punjai*s without any irrigation facilities.[13]
3. Two-thirds to the landlord and one-third to the tenant. The land in this type are *nanjai*s (wet lands) in a dry environ which are irrigated from *eri*s (tanks) and tubewells lifted by diesel engines or electric motors. In this type, the landlord and his tenant bear two-thirds and one-third of the costs for manure and fertilizers and pesticides respectively. Besides, the landlord alone pays for the maintenance charges on the irrigation facilities (such as those for engine oil, electric current, repairing the installed machines, etc.) and cartage; and the tenancy only bears the cost for seeds, operation of the irrigation facilities (usually 3-4 hours per day in the season), and a series of agricultural work.
4. Five-sixths to the landlord and one-sixth to the tenant. As in the third type, the lands are *nanjai*s irrigated by *kavalai*s. The landlord and his tenant equally bear the *kavalai* work and much of the agricultural work. The landlord alone pays for manure and fertilizers, pesticides, and seeds.
5. Seven-eighths to the landlord and one-eighth to the tenant. The physical conditions of the lands are the same as the third and fourth types. The landlord and his tenant equally share the operational work for irrigation and most of the necessary agricultural work. The landlord alone pays for manure and fertilizers, pesticides, seeds, cartage, and the maintenance charges on the irrigation facilities.

Cultures of Asia and Africa [ISLCAA], 1981), pp. 97-164. And also see Mizushima's recent article, "Changes, Chances and Choices: The Perspectives of Indian Villagers," *Socio-cultural Change in Villages in Tiruchirapalli District, Tamilnadu, India, Part 2 (Modern Period), no. 1* (Tokyo: ISLCAA, 1984), pp. 115-116.

12. Shanmugam P. Subbiah, "Rural Base in a South Indian Village: A Study into Its Stuctural and Spatial Patterns in Mahizambadi Village in Tamil Nadu," *Studies in Socio-Cultural Change in Rural Villages in Tiruchirapalli District, Tamil Nadu, India,* no. 4 (Tokyo: ISLCAA, August 1981), p. 66.

13. Ibid., p. 66.

The land-tax in *varam* is paid by the landowner in Lalgudi Taluk of Tiruchy District.[14] The above examples of share-ratios reveal a basic principle: the partner who bears the greater cost gets the greater share of the gross produce, depending, of course, on the physiographic conditions of the *varam* lands to which the different modes of irrigation are employed.

In recent years in India, especially since the 1960s, irrigation systems have developed to a great extent, although there are still regional variations. The development of irrigation systems is one of the key factors responsible for the so-called "Green Revolution." This is true at the *taluk* level of Tiruchy District. In the dry villages of Lalgudi Taluk, modern irrigation methods have been emphasized by innovative landowners: (1) many deep tubewells with electric motors have been installed in new places, (2) some of the traditional wells[15] have been converted into motor-pumped tubewells, and (3) quite a few low-lift pumps have been introduced, replacing the traditional *aetram-eravai*s and *kavalai*s to some extent. In a dry environment the farming of *nanjai*s (wet lands) with the new irrigation systems, accompanied by modern agricultural inputs, provides much higher yields than the *punjai*s (nonirrigated dry lands) or *nanjai*s using the traditional irrigation systems, although it requires more capital and a larger labor force. In accordance with the rapidly increasing capital-labor trend, careful farm management has become a more important factor in successful farming.

There were various responses from the *varam* landowners about the general trend toward the "modernization of farming" in Lalgudi Taluk. Some innovative landowners have cancelled the *varam* contracts and cultivate the land under their own management. Other landowners have converted from the *varam*s to the *kuttagai*s after the installment of oil engines and electric motors.[16] The landowners presently involved in *varam*s are interested, to a varied extent, in the managerial aspects of farming, along with bearing more of the costs for modern inputs, so that they can realize a much higher profit per unit of land than under traditional farming. It should be noted, however, that some of the above landowners have started

---

14. In some areas in Chingleput District, the land tax is paid by the tenant. See Mencher, op. cit., p. 84.

15. The traditional wells are simply hand-dugged and are 5-10 meters in depth and 5-8 meters in diameter, or if square, 7-10 meters in length without pipes.

16. Mizushima, "Change, Chances, and Choices: The Perspectives of Indian Villagers," op. cit., p. 116.

the *varam* contracts even after the installment of the energized irrigation facilities in 1970s. Thus, the crucial question arises: "What are the reasons for the landowners to keep the *varam* contracts?"

The basic reason is the availability of the agricultural labor force at two levels: (1) the working member(s) allocated to farming within an individual household, and (2) the wage laborers and *pannaiyals* (agricultural laborers on a yearly basis) in the villages or regions, although the two components work out in an integrated fashion. The landowners' varied labor situations are primarily responsible for the decision of whether to enter the *varam* tenure, and the degree of their involvement in the *varam* farming when the decision is made. As indicated earlier, the "modernization of farming" has induced a great demand on the labor force. However, the demand is higher in the wet villages than in the dry villages. Moreover, the laborer's wage in the wet villages is at least 1.5 times higher than that in the dry. Consequently, a great number of wage laborers in the dry villages move to the wet villages in season, and, by extension, there is a seasonal scarcity of wage laborers in the dry villages. Therefore, the landowners whose farming largely depends upon an outside labor force have to seek some way to obtain a year-around stable labor force. One way to solve the problem for such landowners is to choose the *varam* tenancy by providing some reasonable conditions for their tenants.

As indicated before, the landowners are afraid that their tenants will gain a "permanent right of cultivation" on the leased lands under the 1969 Tenancy Act. For this reason, the landowners generally prefer the *varam* to the *kuttagai,* because the *varam* landowners can defend themselves from the tenants' claims, should they happen, by insisting that the landowners have been managing the farming.

Land Tenure in Peruvalanallur

Areas and Households Involved in Types of Land Tenure

The intra-village and inter-village land tenure practices should be seen in relation to the general landholding pattern of the village. As indicated already in the previous section, only the *kuttagai* and *otti* types of tenure are observed and dominantly practices in Peruvalanallur. It will be recalled from the discussion at the beginning of the chapter that *kuttagai* is a fixed rent form of tenure and that *otti* is an usufructuary mortgage tenure in which the tenant by depositing a certain amount of cash in advance with the landowner secures the rights to the produce of the land for a stated period, after which the cash advance is returned without any payment of interest.

Table 14 shows the basic statistics of the involved areas (with a distinction between wet and dry lands), between land leased out by landowners or leased in by tenants, between land located in Peruvalanallur or in other villages, and between land in the *kuttagai* and *otti* systems. As indicated already in chapter IV, in 1979-80 the Peruvalanallur villagers owned 1,656.63 acres consisting of 681.29 acres (wet: 582.14 acres; dry: 99.15 acres) in Peruvalanallur (tables 9 and 10), and 975.34 acres in the other villages (wet: 279.67 acres; dry: 695.67 acres). As shown in table 14 of these total lands owned by Peruvalanallur villagers, 18 percent, or 300.27 acres were leased, 205.24 acres under *kuttagai* (line 1 last column), and 95.03 acres under *otti* (line 11 last column). Including both *kuttagai* and *otti* the leased areas consisted of 212.84 acres of wet land and 87.43 acres of dry land.

The Hindu temples and other religious organizations of Peruvalanallur (line 2) owned 37.25 acres (wet: 36.28 acres; dry: 0.97 acres) of cultivable lands within the village territory. Besides, some other Hindu temples which belonged to four other villages (line 5) also owned 9.80 acres of the wet land in Peruvalanallur. Most of these lands were leased to the Peruvalanallur tenants under the *kuttagai* tenancy.

On the other hand, the other villages and town dwellers (excluding temples) in 1979-80 owned 191.41 acres (wet and dry lands) in Peruvalanallur (table 8). Out of this area, 27.84 acres of wet land were leased to the Peruvalanallur tenants under the *kuttagai* or *otti* tenancy. Besides, some landowners in the other villages had the *kuttagai* and *otti* tenures of 28.19 acres (wet: 27.63 acres; dry: 0.56 acres) with the Peruvalanallur tenants in which the involved lands were located outside Peruvalanallur (but mostly in the nearby villages).

Figure 21 shows the spatial distribution of the *kuttagai* and *otti* lands in Peruvalanallur. These leased lands are dispersed fairly evenly through the wet lands. They constituted some 23 percent of the wet lands, or 191.12 acres (total of lines 7 and 13, first column) out of a total of 825.38 acres of agricultural wet land (table 14). But of the 122.13 acres of agricultural dry lands, only 6.73 acres or just over 5 percent were leased. Thus of the arable land of the village a substantial fraction of the wet lands but a negligible fraction of the dry lands are cultivated not by the

TABLE 14

<u>KUTTAGAI</u> AND <u>OTTI</u>: BASIC STATISTICS OF AREAS LEASED-OUT AND -IN BY PERUVALANALLUR VILLAGERS OF LALGUDI TALUK, TIRUCHIRAPPALLI DISTRICT, TAMIL NADU, INDIA (1979-80)

(Unit: in Acre)

		Location of Lands in: Peruvalanallur			Location of Lands in: Other Villages			Total		
		Wet	Dry	Total	Wet	Dry	Total	Wet	Dry	Total
I. Kuttagai	Peruvalanallur:									
	1. Villagers	48.31	4.91	53.22	83.73	68.29	152.02	132.04	73.20	205.24
	2. Temples & Others	36.28	0.97	37.25	-	-	-	36.28	0.97	37.25
	3. (1 + 2)	84.59	5.88	90.47	83.73	68.29	152.02	168.32	74.17	242.49
A. Landowners	Other Villages:									
	4. Villagers	21.75	-	21.75	22.51	0.06	22.57	44.26	0.06	44.32
	5. Temples	9.80	-	9.80	-	-	-	9.80	-	9.80
	6. (4 + 5)	31.55	-	31.55	22.51	0.06	22.57	54.06	0.06	54.12
	7. (3 + 6)	116.14	5.88	122.02	106.24	68.35	174.59	222.38	74.23	296.61
B. Tenants	8. Peru. Villagers	112.20	5.88	118.08	26.70	0.50	27.20	138.90	6.38	145.28
	9. Other Villagers	3.94	-	3.94	79.54	67.85	147.39	83.48	67.85	151.33
	10. (8 + 9)	116.14	5.88	122.02	106.24	68.35	174.59	222.38	74.23	296.61
II. Otti	11. Peru. Villagers	68.89	0.85	69.74	11.91	13.38	25.29	80.80	14.23	95.03
	12. Other Villagers	6.09	-	6.09	5.12	0.50	5.62	11.21	0.50	11.71
A. Landowners	13. (11 + 12)	74.98	0.85	75.83	17.03	13.88	30.91	92.01	14.73	106.74
	14. Peru. Villagers	70.01	0.85	70.86	15.16	5.88	21.04	85.17	6.73	91.90
B. Tenants	15. Other Villagers	4.97	-	4.97	1.87	8.00	9.87	6.84	8.00	14.84
	16. (14 + 15)	74.98	0.85	75.83	17.03	13.88	30.91	92.01	14.73	106.74
III. Kuttagai and Otti	17. Areas leased-out by Peruvalanallur Villagers, Temples, and Others: (3 + 11)	153.48	6.73	160.21	95.64	81.67	177.31	249.12	88.40	337.52
	18. Areas leased to Peruvalanallur Tenants: (8 + 14)	182.21	6.73	188.94	41.86	6.38	48.24	224.07	13.11	237.18

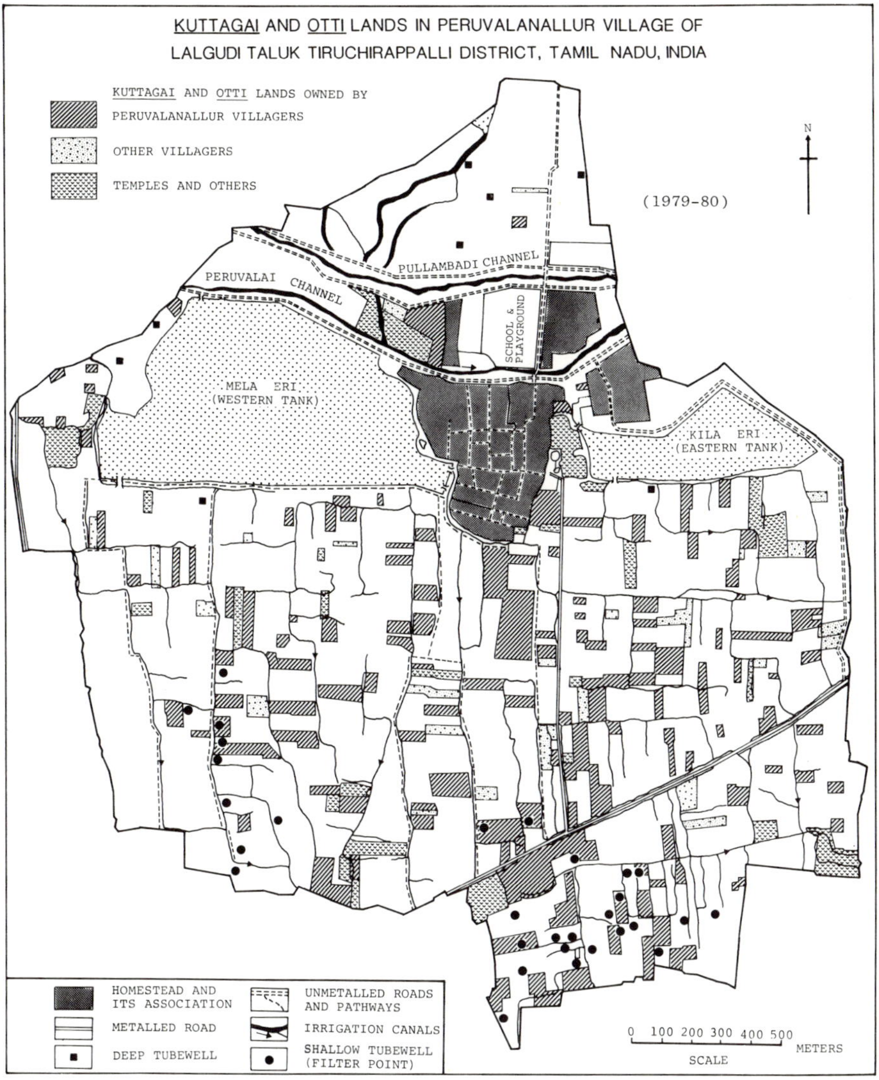
KUTTAGAI AND OTTI LANDS IN PERUVALANALLUR VILLAGE OF
LALGUDI TALUK TIRUCHIRAPPALLI DISTRICT, TAMIL NADU, INDIA
KUTTAGAI AND OTTI LANDS OWNED BY
PERUVALANALLUR VILLAGERS
OTHER VILLAGERS
TEMPLES AND OTHERS
(1979-80)
N
PULLAMBADI CHANNEL
PERUVALAI CHANNEL
SCHOOL & PLAYGROUND
MELA ERI
(WESTERN TANK)
KILA ERI
(EASTERN TANK)
HOMESTEAD AND ITS ASSOCIATION
METALLED ROAD
DEEP TUBEWELL
UNMETALLED ROADS AND PATHWAYS
IRRIGATION CANALS
SHALLOW TUBEWELL (FILTER POINT)
0 100 200 300 400 500
METERS
SCALE

Fig. 21

landowners but by tenants. Moreover, we should realize that the Peruvalanallur villagers owned more of the leased lands outside the village than within the village (177.31 acres compared with 160.21 acres, table 14, line 17). On the other hand, of the land leased by Peruvalanallur villagers, most was physically within the village (188.94 acres) rather than outside the village (48.24 acres, line 18 in table 14).

Next, we are concerned with the involvement of households in the types of land tenure in the studied village. Although 310 households or 35.35 percent of the village total (874 households) in 1979-80 were involved in the two types of the land tenures (the *kuttagai* and *otti),* their involvements are very complex as shown in table 15. The numbers of households for the *kuttagai* were 165 households (landowner side only: 36; tenant side only: 128; and both sides: 86; tenant side only: 116; and both sides: 8). Of these households (165 for the *kuttagai* and 210 for the *otti),* 65 households practiced both types of tenancies.

The Kuttagai

Of the 874 households in Peruvalanallur in 1979-80, 165 households or 19 percent of the village total were involved in the *kuttagai* tenancy, of which 37 households were landowners and 129 were tenants (table 15), although one household was involved on both sides.[17] Both landowners and tenants of Peruvalanallur had *kuttagai* transactions not only with their own villagers but also with the other villagers (tables 16 and 17 and figs. 22, 23 and 24).

The 37 households leased 205.24 acres (wet: 132.04 acres; dry: 73.20 acres) under *kuttagai* tenancy or 12 percent of the total area (table 16). Of these *kuttagai* lands, there were 53.22 acres (wet: 48.31 acres; dry: 4.91 acres) distributed within Peruvalanallur and 152.02 acres (wet: 83.73 acres; dry: 68.29 acres) outside the village (table 14, line 3 and figs. 23 and 24). Of these available *kuttagai* lands leased by the Peruvalanallur villagers, the tenants of the same village cultivated 50.63 acres (wet: 45.72 acres; dry: 4.91 acres) located in their own village teritory, and 4.63 acres (wet: 4.19 acres; dry: 0.44 acre) in the other villages (fig. 24); while the tenants in 19 other villages cultivated the remaining 151.33 acres (wet: 83.48 acres; dry: 67.85 acres) which, except for 2.59 acres of the wet land in Peuvalanallur, were located in the respective tenants' villages or their nearby villages (table 17, part I).

---

17. This case in the *kuttagai* tenancy is unusual, but is understandable when we know that the household leased the land from the Hindu temple.

TABLE 15

KUTTAGAI AND OTTI: NUMBER OF HOUSEHOLDS AS OWNERS OR TENANTS IN PERUVALANALLUR, TIRUCHIRAPPALLI DISTRICT, TAMIL NADU, INDIA (1979-80)

Sign (X): applicable   Sign(0): not applicable

	Kuttagai		Otti		
	Landowner-side	Tenant-side	Landowner-side	Tenant-side	No. of Households
	X	0	0	0	29
	0	X	0	0	70
	0	0	X	0	49
	0	0	0	X	92
	X	X	0	0	1
	0	0	X	X	4
	X	0	X	0	4
	X	0	0	X	2
	0	X	X	0	33
	0	X	0	X	22
	0	X	X	X	3
	X	0	X	X	1
Total	37	129	94	124	310

Notes: 1. The figures for each column and row show the total number of households in that type of involvement.

TABLE 16

KUTTAGAI TRANSACTIONS BY CASTE IN PERUVALANALLUR VILLAGE OF LALGUDI TALUK, TIRUCHIRAPPALLI DISTRICT, TAMIL NADU, INDIA (1979-80)

Unit: in Acre

I. Landowner Side

Castes	No. of Households	Areas leased-out		
		Wet	Dry	Total
Reddiar	16	116.54	57.79	174.33
Udaiyar	6	5.35	3.50	8.85
Gounder	4	0.68	7.50	8.18
Muslim	6	6.45	2.91	9.36
Achari	1	1.00	-	1.00
Mooppanar	1	1.00	-	1.00
Protestant	1	-	1.50	1.50
Hindu Pallan	1	0.42	-	0.42
Domban	1	0.60	-	0.60
	37	132.04	73.20	205.24
Hindu Temples		35.97	-	35.97
Mosque		0.31	-	0.31
Catholic Church		-	0.97	0.97
		36.28	0.97	37.25
Other Villagers		44.26	0.06	44.32
Hindu Temples belonged to Other Villages		9.80	-	9.80
		54.06	0.06	54.12
TOTAL		222.38	74.23	296.61

II. Tenant Side

Castes	No. of Households	Areas leased-in		
		Wet	Dry	Total
Reddiar	25	43.68	0.50	44.18
Udaiyar	11	12.06	-	12.06
Gounder	9	10.60	2.97	13.57
Muslim	1	2.16	-	2.16
Nadar	1	0.28	-	0.28
Muthuraja	8	12.79	-	12.79
[illegible]ariyari	2	0.90	-	0.90
Pandaram	4	2.91	-	2.91
Hindu Pallan	53	41.66	1.94	43.60
Hindu Parayan	1	1.00	-	1.00
Catholic Pallan	7	5.55	-	5.55
Catholic Parayan	7	5.31	0.97	6.28
	129	138.90	6.38	145.28
Other Villagers		83.48	67.85	151.33
TOTAL		222.38	74.23	296.61

TABLE 17

KUTTAGAI TRANSACTIONS BETWEEN PERUVALANALLUR AND OTHER VILLAGERS IN LALGUDI TALUK, TIRUCHIRAPPALLI DISTRICT, TAMIL NADU, INDIA (1979-80)

Unit: in Acre

I. Peruvalanallur Landowners leased-out to Other Villagers

Village Names of Tenants	Areas leased-in		
	Wet	Dry	Total
Vengangudi(#19)	3.75	-	3.75
Appadurai(#23)	-	3.50	3.50
Esanakkorai(#24)	-	2.00	2.00
Valadi(#26)	2.00	-	2.00
Sirumarudur(#27)	8.61	-	8.61
V. Turaiyur(#28)	23.48	-	23.48
S. Kannanur(#29)	3.87	-	3.87
Marudur (#30)	3.77	-	3.77
R. Valavanur(#31)	13.86	0.95	14.81
Kumulur(#37)	1.15	-	1.15
Sirumayangudi(#62)	3.52	-	3.52
Poovalur(#64)	0.70	-	0.70
Sirudaiyur(#71)	0.65	-	0.65
Thirumangalam(#72)	12.58	-	12.58
Neikuppai(#74)	4.10	-	4.10
Siruganur(#91)	-	54.40	54.40
Outside of Lalgudi Taluk:			
Annamangalam (Perambalur TK.)	-	1.50	1.50
Natthampalayam ( )	-	5.50	5.50
Pudukkottai (District Capital)	1.44	-	1.44
Total	83.48	67.85	151.33

Notes:

1. Each number in parentheses corresponds to the revenue village number listed in Table 1.

2. The lands leased-in by the tenants of Sirumayangudi (#62) and Poovalur (#64) belonged to the Hindu temples of Peruvalanallur. Subtracting the temple lands (1.35 acres), the total area leased-out by the Peruvalanallur villagers was 149.98 acres.

TABLE 17 (continued)

Unit: in Acre

II. Peruvalanallur Tenants Leased-in from Other Villagers

Village Names of Landowners	Areas leased-out		
	Wet	Dry	Total
Thachankuruchi(#34)	0.74	-	0.74
Reddimangudi(#35)	0.43	-	0.43
Kumulur(#37)	0.28	-	0.28
Peruvalapur(#42)	2.68	0.06	2.74
Kanakkiliyanallur(#43)	2.92	-	2.92
Pullambadi(#45)	1.99	-	1.99
Pallapuram(#63)	0.30	-	0.30
Poovalur(#64)	5.58	-	5.58
Sirudaiyur(#71)	1.68	-	1.68
Perakambi(#86)	1.08	-	1.08
Neykulam(#89)	1.30	-	1.30
Siruganur(#91)	3.97	-	3.97
Garudamangalam(#104)	2.03	-	2.03
Kallakudi(#112)	1.00	-	1.00
Outside of Lalgudi Taluk:			
Kalpadi (Perambalur TK.)	0.74	-	0.74
Thiruvanaikoil (Trichy TK.)	8.03	-	8.03
Tiruchy Town	2.88	-	2.88
Thathaiyankarpet (Musiri TK.)	4.02	-	4.02
Erumaipatti (Namakkal TK., Salem Dt.)	2.16	-	2.16
Neiveli (S. Arcot Dt.)	0.45	-	0.45
	44.26	0.06	44.32
Temples:			
Pullambadi(#45)	1.00	-	1.00
Poovalur(#64)	2.94	-	2.94
Siruganur(#91)	4.38	-	4.38
Uttathur(#116)	1.48	-	1.48
	9.80		9.80
Total	54.06	0.06	54.12

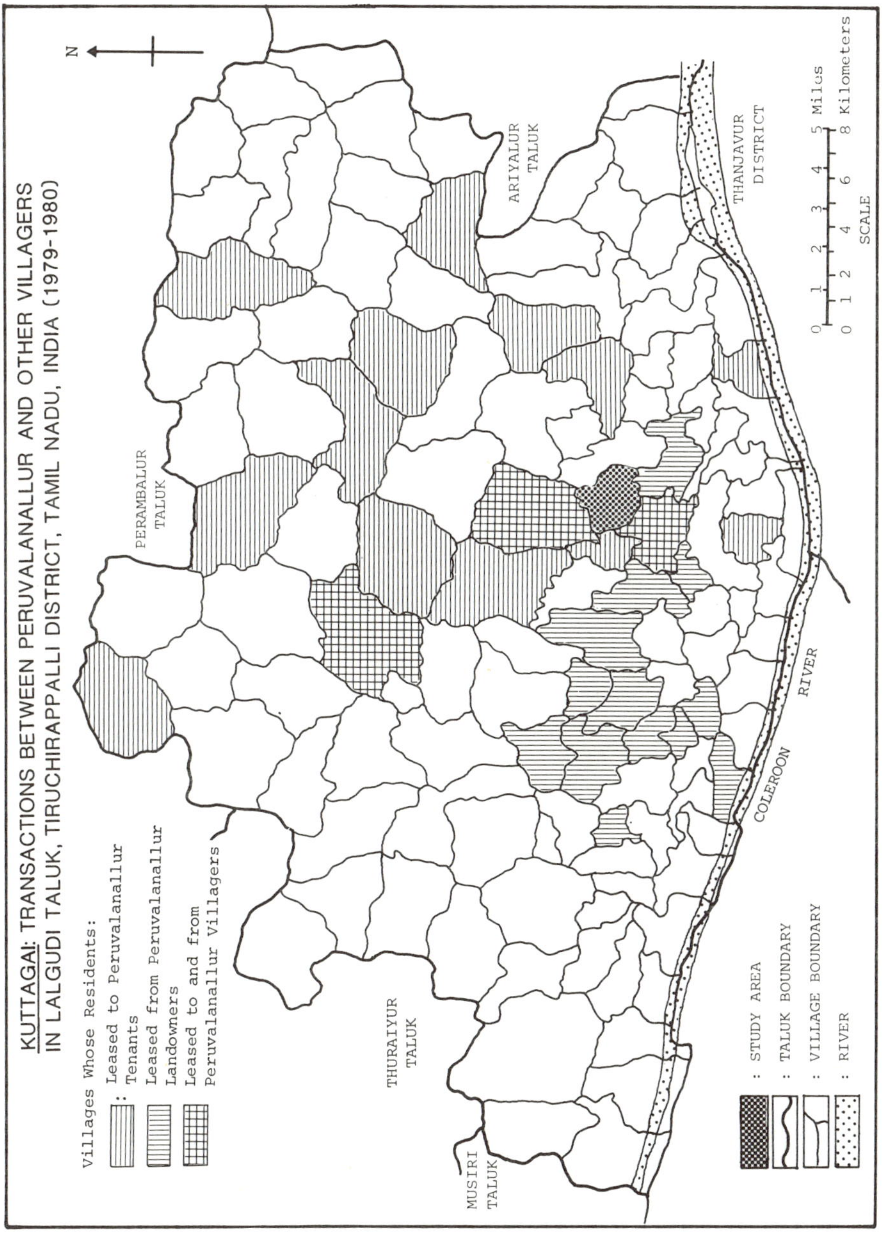
KUTTAGAI: TRANSACTIONS BETWEEN PERUVALANALLUR AND OTHER VILLAGERS
IN LALGUDI TALUK, TIRUCHIRAPPALLI DISTRICT, TAMIL NADU, INDIA (1979-1980)
N
Villages Whose Residents:
: Leased to Peruvalanallur Tenants
Leased from Peruvalanallur Landowners
Leased to and from Peruvalanallur Villagers
PERAMBALUR TALUK
ARIYALUR TALUK
THANJAVUR DISTRICT
THURAIYUR TALUK
MUSIRI TALUK
COLEROON
RIVER
: STUDY AREA
: TALUK BOUNDARY
: VILLAGE BOUNDARY
: RIVER
0 1 2 3 4 5 Miles
0 1 2 4 6 8 Kilometers
SCALE

Fig. 22

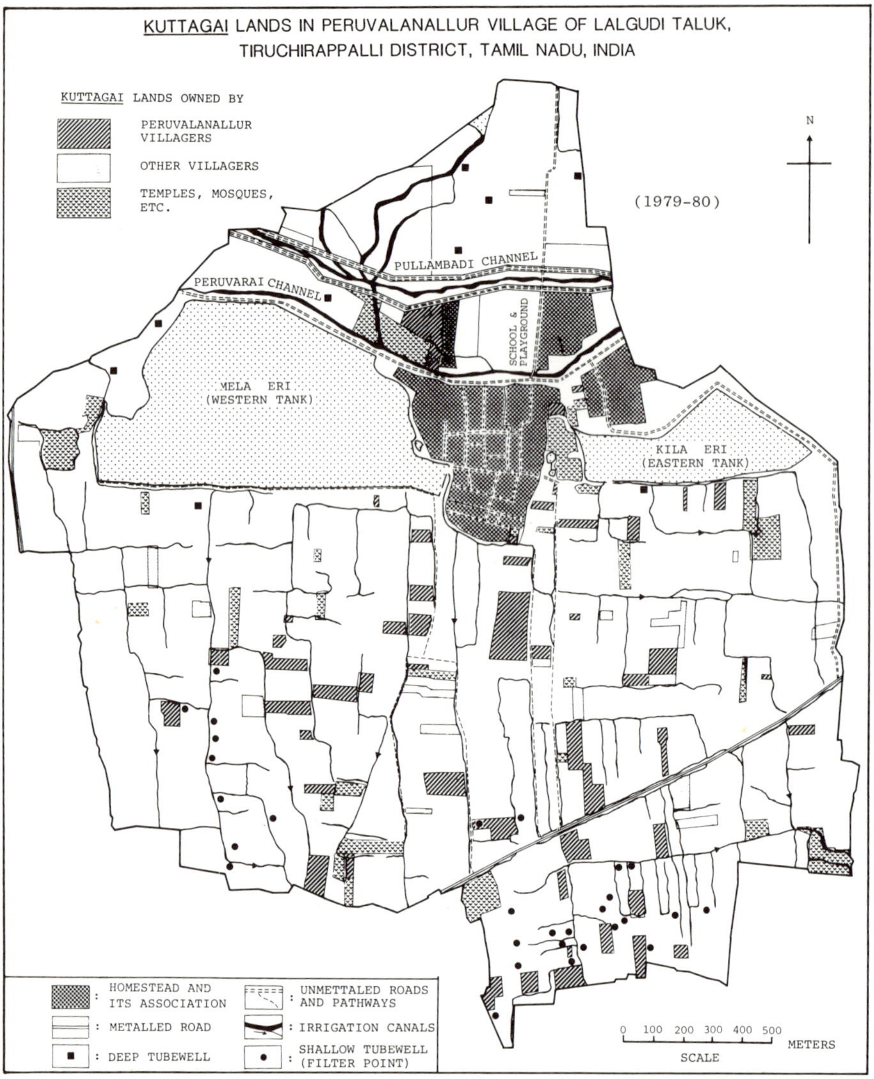
KUTTAGAI LANDS IN PERUVALANALLUR VILLAGE OF LALGUDI TALUK,
TIRUCHIRAPPALLI DISTRICT, TAMIL NADU, INDIA
KUTTAGAI LANDS OWNED BY
PERUVALANALLUR VILLAGERS
OTHER VILLAGERS
TEMPLES, MOSQUES, ETC.
(1979-80)
N
PULLAMBADI CHANNEL
PERUVARAI CHANNEL
SCHOOL & PLAYGROUND
MELA ERI (WESTERN TANK)
KILA ERI (EASTERN TANK)
: HOMESTEAD AND ITS ASSOCIATION
: METALLED ROAD
: DEEP TUBEWELL
: UNMETTALED ROADS AND PATHWAYS
: IRRIGATION CANALS
: SHALLOW TUBEWELL (FILTER POINT)
0 100 200 300 400 500
METERS
SCALE

Fig. 23

KUTTAGAI: AREAS LEASED-OUT AND -IN BY PERUVALANALLUR VILLAGERS OF LALGUDI TALUK, TIRUCHIRAPPALLI DISTRICT TAMIL NADU, INDIA (1979-80)

(Unit: in Acre)

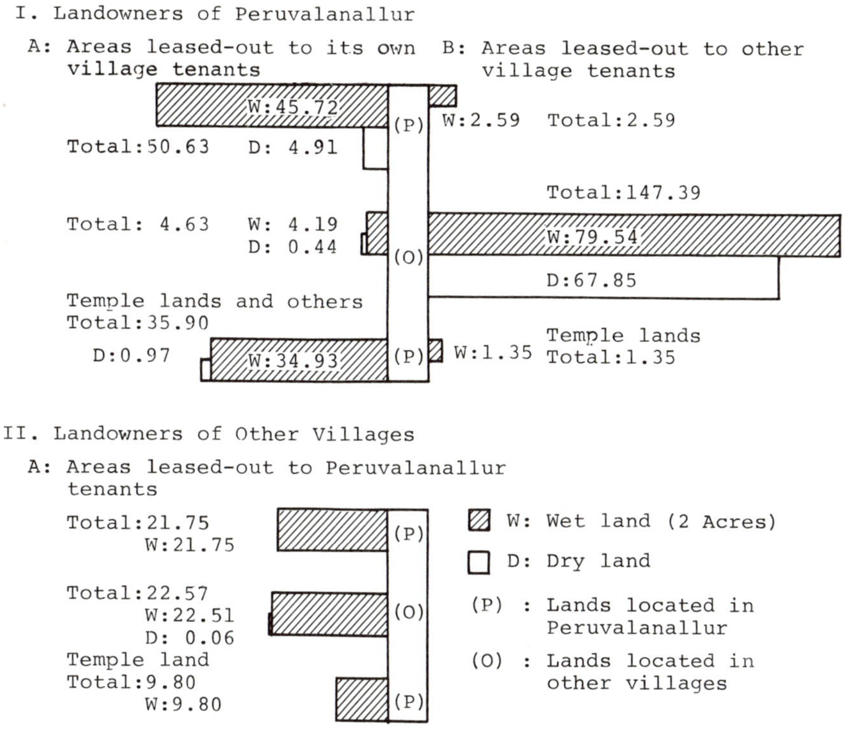

Fig. 24

On the other hand, the Peruvalanallur tenants leased 44.32 acres (wet: 44.26 acres; dry: 0.06 acre) from landowners who lived in 21 other villages and towns (table 17, part II). Of these lands, 21.75 acres (wet land only) were located within Peruvalanallur and 22.57 acres (wet land: 22.51 acres; dry: 0.06 acre) were outside the village (fig. 24). As indicated already, the arable lands belonging to the Hindu temples and other religious organizations are usually leased under the *kuttagai* system. These lands in Peruvalanallur were cultivated mostly by its own village tenants.

To summarize, 129 tenant households of the studied village in 1979-80 leased 145.28 acres (wet: 138.90 acres; dry: 6.38 acres) under *kuttagai* not only from the landowners of their own village but also from those in the nerby villages, including the Hindu temples and other religious organizations (table 14, line 8).

The intra- and inter-village land tenure transactions should be discussed by taking into account conditions from the sides of both the landowner and tenant. The spatial distribution of *kuttagai* areas in and outside Peruvalanallur and leased by the village landowners are related primarily to the fact that the Peruvalanallur villagers own more acreage in the other villages (975.34 acres) than in their own village (681.29 acres) and that the landowners usually try to cultivate the lands located near their homesteads.

It should be noted that the Peruvalanallur tenants have been overwhelmingly interested in cultivating wet lands rather than dry lands and that the lands were located mostly in and around the village. In this respect, the distance factor and the different productivities or yields per unit area of crops between the wet and dry lands are thought to be responsible. It is generally true that the available *kuttagai* lands in a given village, regardless of their ownership, were mostly cultivated by the tenants of the same village or by the neighboring villagers. This indicates that the distance factor is important in the practical operation of cultivating *kuttagai* lands. The fact that the Peruvalanallur tenants have actually leased wet lands almost exclusively can be explained by this factor too, since the village itself lies in an extensive region with a wet environment. However, the distance factor depends upon other agricultural conditions in a given village or area.

The yield of crops per unit of area in the wet land in a wet environment like Peruvalanallur is generally more than three times higher than that in the dry land, although there is a great variation within each of the wet and dry lands in a given milieu. This is another reason why most of the Peruvalanallur tenants desire to cultivate wet rather than dry lands whenever possible.

Next we will examine the households involved in the *kuttagai* tenancy in relation to the caste and size of landholding. The basic statistics of the involved households and areas by caste groups are already shown in table 16. More detailed statistics are shown in figure 25 in which all the cases of *kuttagai* transactions between the landowners and tenants are arranged under each caste group. Both table 18 and figure 26 show the distribution of the involved households and areas by the size of landholding.

Of the 33 caste groups in Peruvalanallur, there were only 16 groups involved in the *kuttagai* tenancy (landowner side only: 4 groups; tenant side only: 7 groups; and both sides: 5 groups). Among the 9 caste groups of the *kuttagai* landowners, the *Reddiars* (16 households) leased 174.33 acres or 85 percent of the total *kuttagai* lands (205.24 acres) owned by the Peruvalanallur villagers (table 16). The other caste groups of the *kuttagai* landowners who leased sizeable areas were the Muslims (9.36 acres), the *Udaiyars* (8.85 acres), and the *Gounders* (8.18 acres). Each of the remaining 5 caste groups respectively had one houehold, and its leased land was a very small area (1.50 acres at most). On the other hand, among the 12 caste groups of the *kuttagai* tenants, 5 groups (the *Reddiars*, *Udaiyars*, *Gounders*, *Muthurajas*, and Hindu *Pallans*) were the most important ones as far as the involved households and the involved households and their leased areas are concerned.

As far as the number of the involved landowners' households in the *kuttagai* tenancy is concerned, there was no relation to the size of landholding. Unlike our supposition, there were 17 households belonging to the marginal landholding group with 10 households under 1 acre and 7 households 1-2 acres (table 18). Most of these households leased all of their lands.[18] However, since their respective sizes of landholding were small by definition, the total leased area and area per household were accordingly small. There is a general trend that the larger landowning households leased more area per household (table 18, last column). In this respect, however, it should be noted that the largest landowning household *(Reddiar)* in the studied village alone leased as much as 136.41 acres of their land (wet: 81.06 acres; dry: 55.35 acres). More specifically, one household alone leased 12.24 acres (wet land only) in 18 different contracts with the Peruvalanallur villagers (who belonged to 7 different caste groups), and 124.17 acres (wet: 68.82 acres; dry: 55.35 acres) in 67 different contracts with the other

---

18. Thus, we may automatically suppose that all of these households had no operational lands of their own. However, some of the households got the leased lands from others under the *otti* tenancy.

RELATIONSHIPS AMONG LANDOWNERS AND TENANTS UNDER KUTTAGAI BY CASTE IN PERUVALANALLUR VILLAGE, LALGUDI TALUK, TIRUCHIRAPPALLI DISTRICT, TAMIL NADU, INDIA (1979 - 1980)

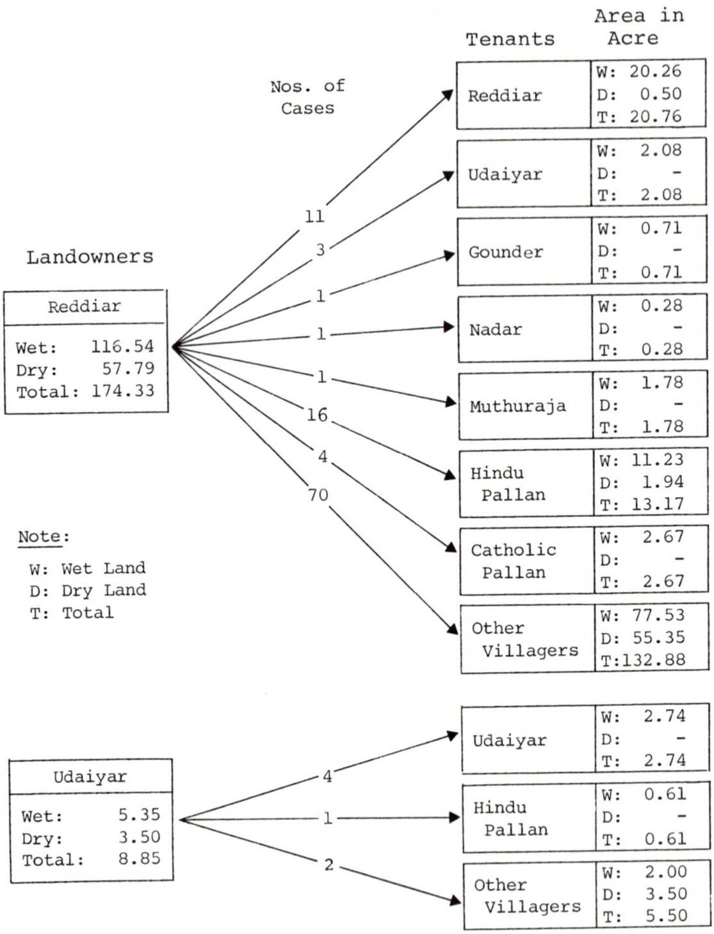

Fig. 25

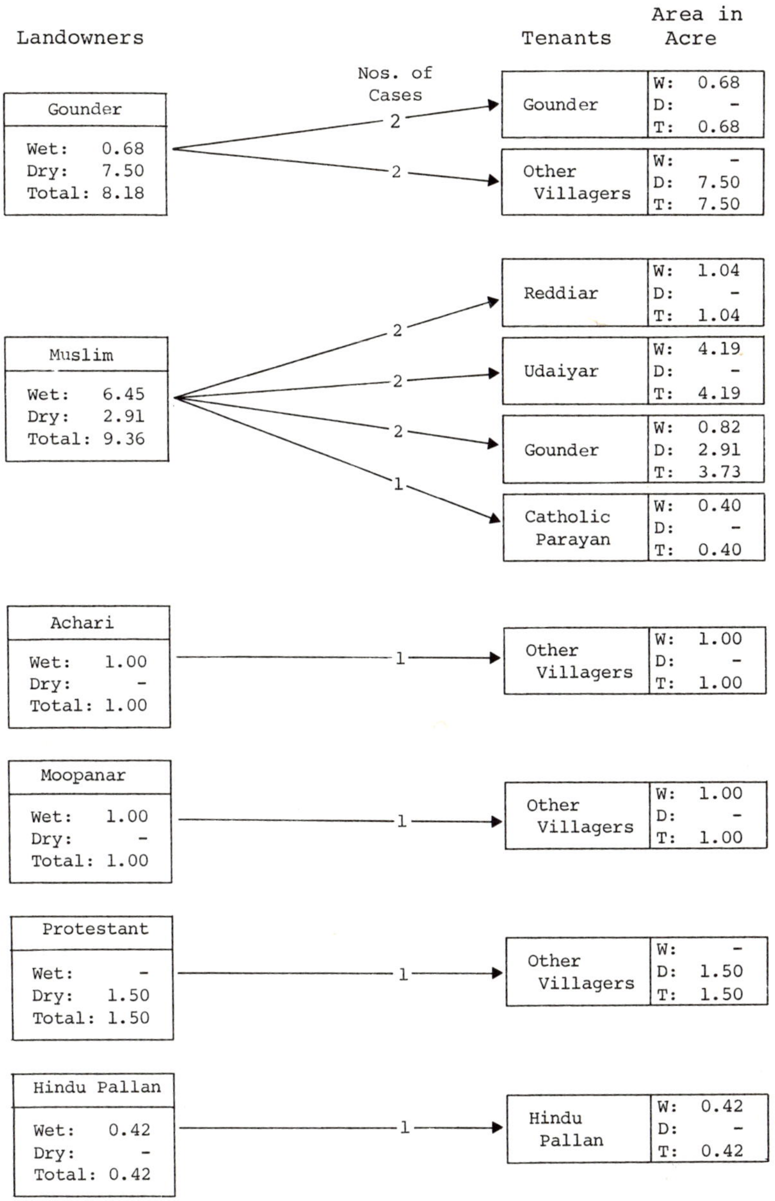
Landowners
Tenants
Area in Acre
Nos. of Cases
Gounder
Wet: 0.68
Dry: 7.50
Total: 8.18
2
2
Gounder
W: 0.68
D: -
T: 0.68
Other Villagers
W: -
D: 7.50
T: 7.50
Muslim
Wet: 6.45
Dry: 2.91
Total: 9.36
2
2
2
1
Reddiar
W: 1.04
D: -
T: 1.04
Udaiyar
W: 4.19
D: -
T: 4.19
Gounder
W: 0.82
D: 2.91
T: 3.73
Catholic Parayan
W: 0.40
D: -
T: 0.40
Achari
Wet: 1.00
Dry: -
Total: 1.00
1
Other Villagers
W: 1.00
D: -
T: 1.00
Moopanar
Wet: 1.00
Dry: -
Total: 1.00
1
Other Villagers
W: 1.00
D: -
T: 1.00
Protestant
Wet: -
Dry: 1.50
Total: 1.50
1
Other Villagers
W: -
D: 1.50
T: 1.50
Hindu Pallan
Wet: 0.42
Dry: -
Total: 0.42
1
Hindu Pallan
W: 0.42
D: -
T: 0.42

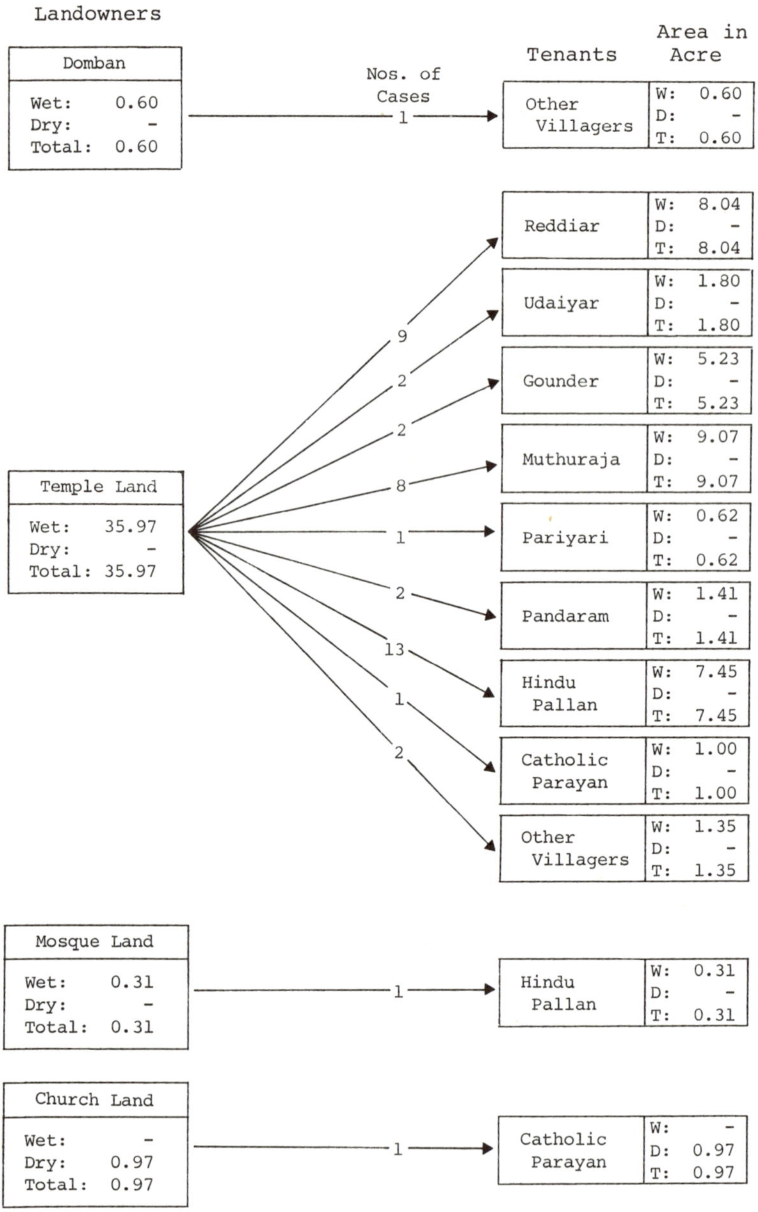
Landowners
Tenants
Area in Acre
Nos. of Cases
Domban
Wet: 0.60
Dry: -
Total: 0.60
1
Other Villagers
W: 0.60
D: -
T: 0.60
Temple Land
Wet: 35.97
Dry: -
Total: 35.97
9
Reddiar
W: 8.04
D: -
T: 8.04
2
Udaiyar
W: 1.80
D: -
T: 1.80
2
Gounder
W: 5.23
D: -
T: 5.23
8
Muthuraja
W: 9.07
D: -
T: 9.07
1
Pariyari
W: 0.62
D: -
T: 0.62
2
Pandaram
W: 1.41
D: -
T: 1.41
13
Hindu Pallan
W: 7.45
D: -
T: 7.45
1
Catholic Parayan
W: 1.00
D: -
T: 1.00
2
Other Villagers
W: 1.35
D: -
T: 1.35
Mosque Land
Wet: 0.31
Dry: -
Total: 0.31
1
Hindu Pallan
W: 0.31
D: -
T: 0.31
Church Land
Wet: -
Dry: 0.97
Total: 0.97
1
Catholic Parayan
W: -
D: 0.97
T: 0.97

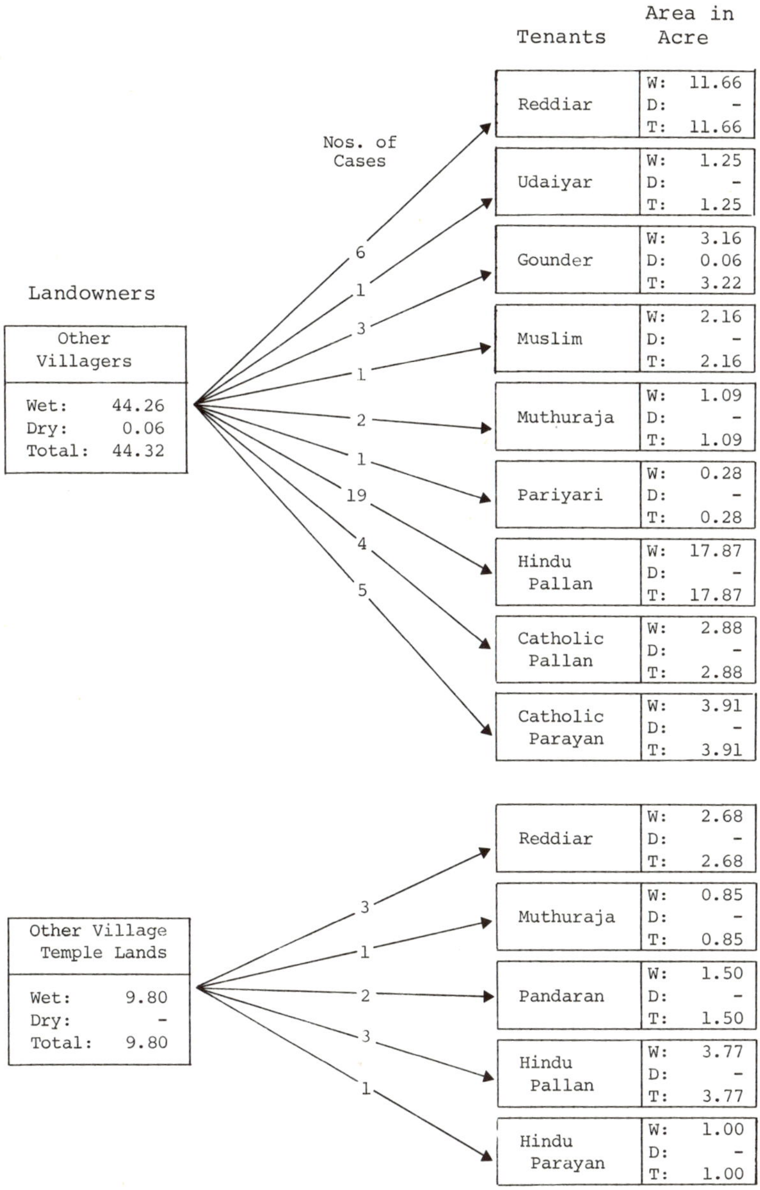
Area in Acre
Tenants
Nos. of Cases
Landowners
Other Villagers
Wet: 44.26
Dry: 0.06
Total: 44.32
6
1
3
1
2
1
19
4
5
Reddiar
W: 11.66
D: -
T: 11.66
Udaiyar
W: 1.25
D: -
T: 1.25
Gounder
W: 3.16
D: 0.06
T: 3.22
Muslim
W: 2.16
D: -
T: 2.16
Muthuraja
W: 1.09
D: -
T: 1.09
Pariyari
W: 0.28
D: -
T: 0.28
Hindu Pallan
W: 17.87
D: -
T: 17.87
Catholic Pallan
W: 2.88
D: -
T: 2.88
Catholic Parayan
W: 3.91
D: -
T: 3.91
Other Village Temple Lands
Wet: 9.80
Dry: -
Total: 9.80
3
1
2
3
1
Reddiar
W: 2.68
D: -
T: 2.68
Muthuraja
W: 0.85
D: -
T: 0.85
Pandaran
W: 1.50
D: -
T: 1.50
Hindu Pallan
W: 3.77
D: -
T: 3.77
Hindu Parayan
W: 1.00
D: -
T: 1.00

TABLE 18

KUTTAGAI: THE INVOLVED HOUSEHOLDS AND AREAS BY SIZE OF LANDHOLDING IN PERUVALANALLUR VILLAGE, LALGUDI TALUK, TIRUCHIRAPPALLI DISTRICT, TAMIL NADU, INDIA (1979-80)

I. Landowner Side (Unit: in Acre)

Categories by Size of Land-holding	No. of House-holds	Wet Land	Dry Land	Total	Area per House-hold
C-1: landless	-	-	-	-	-
C-2: under 1 acre	10	5.59	-	5.59	0.56
C-3: 1-2 acres	7	6.58	1.50	8.08	1.15
C-4: 2-3 acres	5	6.47	2.00	8.47	1.69
C-5: 3-5 acres	4	11.40	3.41	14.81	3.70
C-6: 5-7 acres	1	5.49	-	5.49	5.49
C-7: 7-10 acres	2	9.68	-	9.68	4.84
C-8: 10-15 acres	3	2.00	9.00	11.00	3.67
C-9: 15 & above	5	84.83	57.29	142.12	28.43
Total	37	132.04	73.20	205.24	5.55

II. Tenant Side (Unit: in Acre)

Categories by Size of Land-holding	No. of House-holds	Wet Land	Dry Land	Total	Area per House-hold
C-1: landless	44	30.28	2.91	33.19	0.75
C-2: under 1 acre	32	24.28	-	24.28	0.76
C-3: 1-2 acres	14	19.53	-	19.53	1.40
C-4: 2-3 acres	9	9.03	0.06	9.15	1.02
C-5: 3-5 acres	8	14.84	-	14.84	1.86
C-6: 5-7 acres	6	11.76	0.50	12.26	2.04
C-7: 7-10 acres	6	12.56	-	12.56	2.09
C-8: 10-15 acres	4	8.90	2.91	11.81	2.95
C-9: 15 & above	6	7.66	-	7.66	1.28
Total	129	138.90	6.38	145.28	1.13

KUTTAGAI: PERCENTAGE DISTRIBUTION OF HOUSEHOLDS AND AREAS INVOLVED IN KUTTAGAI BY SIZE OF LANDHOLDING IN PERUVALANALLUR (1979-1980)

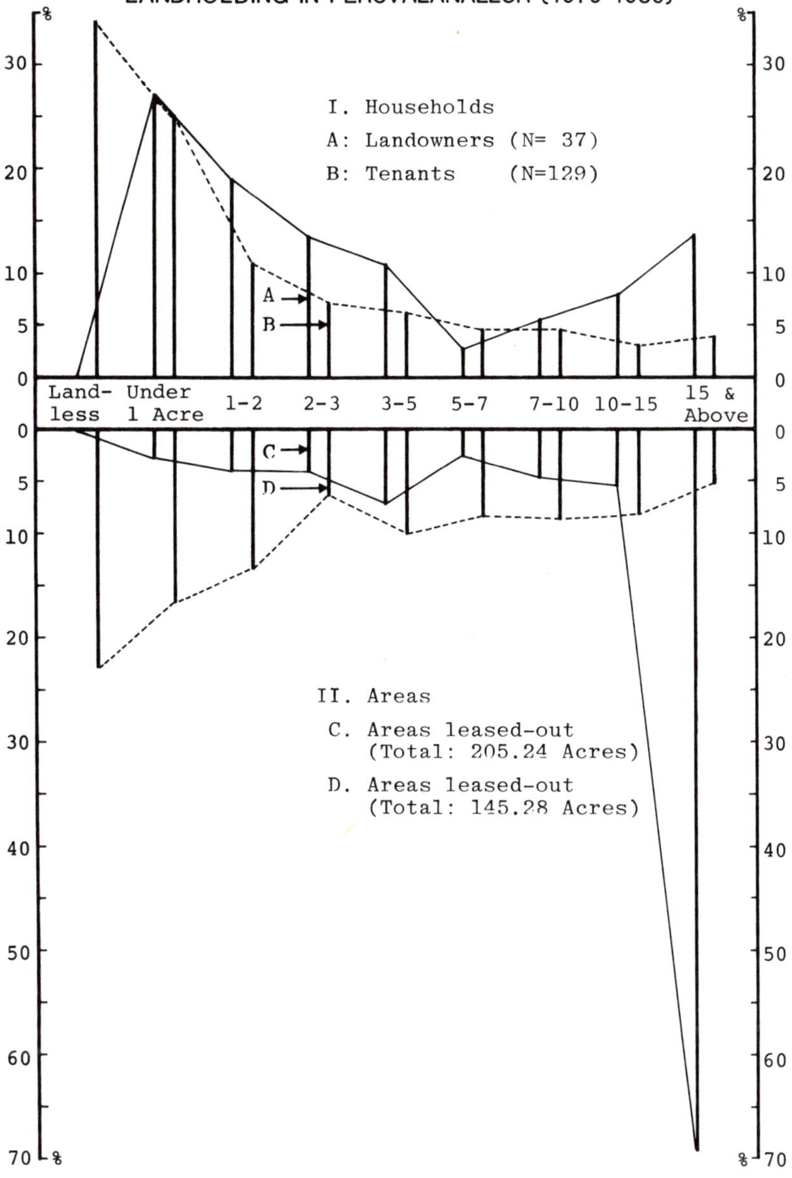

Fig. 26

villagers.[19] This fact is responsible for the extremely high ratio of the leased area by the *Reddiars* in table 16 and figure 25, and by the highest landowning class in table 18 and figure 26 as well.

As would be expected in the *kuttagai* tenancy, the landless and marginal landowning households are more involved in the tenancy. Out of the 129 tenants' households in Peruvalanallur, 90 households or 70 percent consisted of the landless (44 households) and marginal landowning groups with under 2 acres (46 households), and they together leased 77.00 acres or 53 percent of the total *kuttagai* area (145.28 acres) leased by the village tenants. It is generally true that the smaller landowning households leased the smaller extent of area per household. This is certainly related to the tenants' capability for farm management, as the tenants in the *kuttagai* tenancy have to meet all the necessary expenses for the cultivation, although the marginal farmers usually spend a lesser amount for modern agricultural inputs per unit of area than the larger landowning farmers. It is also related to the result of the subdivision of the *kuttagai* lands caused by the "inheritance" practice prevalent among the landless and smaller landowning households.

The *kuttagai* landlords and tenants have reasons for their involvement in that tenancy system. The basic reason for the landlords in the shortage or lack of an agricultural labor force for cultivating the lands by themselves at the time of the contracts. It should be noted, however, that the shortage or lack of the labor force does not necessarily mean that there is no working member(s) within the family, but rather that the families who have working member(s) have them get jobs in the non-agricultural occupations as is indicated in chapter V. If the shortage or lack occurs in the larger landowning households (say, owning more than 5 acres of wet land), they can still maintain a better standard of living from the rent alone. However, in the cases of the marginal landowning households (owning less than 2 acres of wet land), their standard of living is generally low if their income is only from the rent. Regardless of the size of the landholding, there are some landlords who were aged couples or widows (or widowers) without any sons and daughters. Such landlords are generally in a weak position in relation to their tenants. There were two extreme cases where the landlords had not received any rent from their tenants for several years.

---

19. Of the 20 villages whose people leased the *kuttagai* lands from the Peruvalanallur villagers, this landlord owned all the *kuttagai* lands in 5 villages (nos. 27, 30, 72, 74, and 91) and a major part in 3 villages (nos. 28, 29 and 31) listed in table 17.

Table 19 shows the number of counterparts for each landowners and tenants in the studied village. This table reveals that a landowner has a *kuttagai* relationship(s) with one or more tenants, and vice versa for a tenant. The dominant cases were singular counterpart relationships on both sizes (the landowner and tenant). However, there were two exceptional landowners who had a large number of counterparts: one was the biggest landowner in Peruvalanallur who had as many as 85 different counterparts (tenants) in and out of the village (table 19); the other was the Hindu temples who had 40 counterparts.

Now an important question should be asked, "Are there any caste-related contracts in the landlord-tenant partnership in the *kuttagai* tenancy?" Examining each of the contracts, the answer is generally "yes." A landlord usually chooses his tenant(s) either from his own caste or from socio-economically lower-ranking castes compared with his own. Conversely, a tenant seeks his landlord(s) either from his own caste or from socio-economically higher-ranking castes compared with his own (fig. 25). This is certainly related to the barrier of the caste psychology like the one that a *Reddiar* expressed to the author as follows:

> ...although I am a poor farmer, I cannot be the tenant for a *Harijan* landlord. it is impossible for me to be subordinate to any *Harijan*s. If I cultivated the *Harijan*s land, people would laugh at me.

With regard to the above discussion, the one exception is the Muslims who do not have such cultural barriers.

Another important aspect in the landlord-tenant relationship under the *kuttagai* tenancy is that, in a given contract, the size of the landholding of the landlord is not necessarily larger than that of his tenant. This situation is hidden in our tables and figures. The implication of this point is that, under such landlord-tenant relationships, the landlord was not necessarily more influential, socio-economically speaking, than his tenant.

With regard to the landlord-tenant relationships in the *kuttagai* tenancy, it is relevant to raise the question: "How long has the current counterpart (a landlord and his tenant) of the *kuttagai* contract continued?" Figure 27 shows the percentage distribution of the number of current contracts by different periods (in years). This figure reveals that the majority of the current *kuttagai* contracts have been continued for long periods, that is, 80 percent of the total contracts have continued for more than 10 years.

TABLE 19

<u>KUTTAGAI</u>: NUMBER OF COUNTERPARTS FOR EACH LANDOWNER AND TENANT IN PERUVALANALLUR VILLAGE OF LALGUDI TALUK, TIRUCHIRAPPALLI DISTRICT, TAMIL NADU, INDIA (1979-80)

I. Landowner Side		II. Tenant Side	
Number of Counterparts	Number of Cases (Landowners' Households)	Number of Counterparts	Numberof Cases (Tenants' Households)
1 tenant	31 (83.79%)	1 landowner	113 (87.60%)
2 tenants	4 (10.81)	2 landowners	15 (11.62)
6 tenants	1 (2.70)	3 landowners	1 (0.78)
85 tenants	1 (2.70)		
	37 (100.00%)		129 (100.00%)

Note: In addition to the above numbers of the *kuttagai* landowners, Hindu temples and other religious organizations *together* have a considerable number of the tenant counterparts (see Fig. 25).

PERIOD OF CURRENT KUTTAGAI CONTRACTS IN PERUVALANALLUR
LALGUDI TALUK OF TIRUCHIRAPPALLI DISTRICT
TAMIL NADU, INDIA

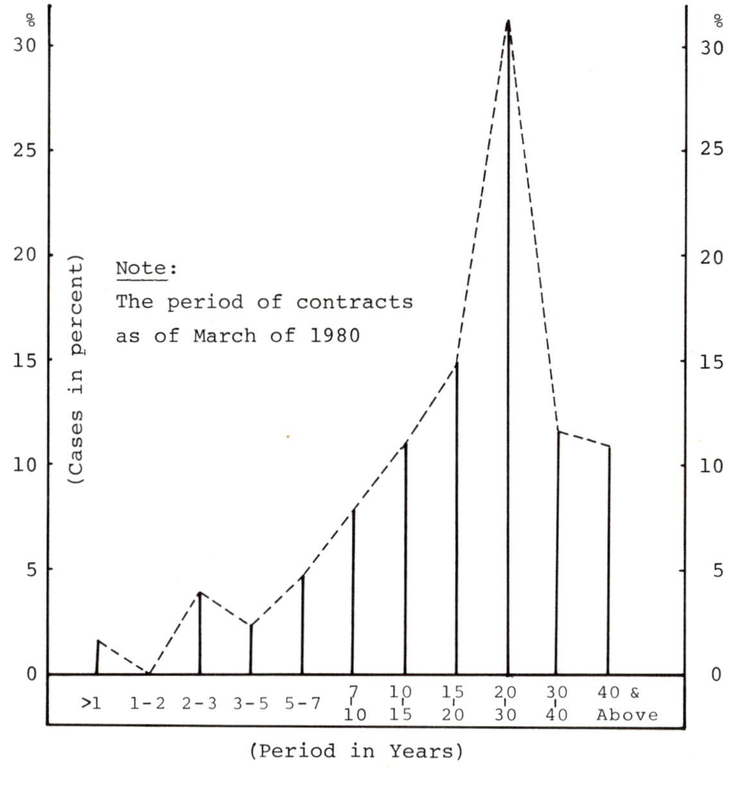

Fig. 27

It should be noted that the Tamil Nadu Agricultural Lands Records of Tenancy Rights Act was passed in 1969. We gathered from, in, and around the studied village that some tenants simply gave up their tenancies; some tried to get tenancies from other landlords; and some were even shifted by the same landlords to another plot. Although we know about these occurrences, we do not know to what extent exactly the above cases actually occurred in relation to the 1969 Act. Nevertheless, the fact that the lower percentage of the recent *kuttagai* contracts which have started within the last 10 years (based on data as of March 1980) is certainly related to the influence of the 1969 Act. It should be noted that many of the recent contracts were those among close kinship groups.

On the other hand, how should we take into consideration the fact that the great majority of the counterparts in the current *kuttagai* contracts has not changed for a long period (in years). One way to look at this is that the landlords might not have been as strict with their tenants as is generally understood, although there have been exceptions. As a rule, if there was a bad harvest season, the tenants would pay a lesser amount of the fixed rent to their landlords, although they are still supposed to pay the remaining part of the rent from the next harvest or at least within a year. However, in practice, this rule has rarely been followed. Thus, a more common practice is that, although the tenants usually pay the full amount of the fixed rent in normal years, some concessions are made on the fixed rent in sub-normal years, depending upon the actual amount of the gross produce from the *kuttagai* lands. Under this practice, the landlords usually do not hastily eject the tenants. Looking at this practice, it should be noted that there are some aspects of share-cropping *(varam* tenure) within the *kuttagai* system.

As indicated already in the previous section, once a *kuttagai* contract is made, there arises the possibility of entering the case in the Tenancy Register. In Peruvalanallur in 1972, the registered areas were 68.67 acres (wet: 61.10 acres; dry: 7.57 acres) for the 100 *kuttagai* contracts, which included not only the lands owned by its own villagers and the Hindu temples (and other religious bodies), but also the lands owned by outside villagers and Hindu temples. It should be noted that, of the above registered areas, the lands which belonged to Hindu temples and other religious bodies in and out of the village occupied a considerable share: 25.68 acres (wet: 18.11 acres; dry: 7.57 acres) in 32 of the contracts. Since we do not have the total leased out area in the village at that time, we cannot get the percentage of the registered leased land.

Even if the contracts are not entered in the Tenancy Register, most of the tenants in Peruvalanallur believe that they have a "permanent" right to cultivate the *kuttagai* lands, although their landlords apparently do not accept the idea. Anyhow, such tenants' belief itself seems to be an important "property" of the *kuttagai* for them, besides enjoying the cultivation of the lands involved.

Then, will these landlord-tenant relationships be continued "permanently"? In answer to this question, the fact is that, by 1980 in Peruvalanallur, the number of cases and the extent of the registered *kuttagai* contracts were reduced to 65 cases and 45.87 acres (wet: 38.30 acres; dry: 7.57 acres), because 14.84 acres (wet land only) in 25 cases had been returned to the landlords mostly with some conditions (receiving some amount of cash money) and the remaining 7.97 acres (wet land only) were purchased by the tenants at reduced prices. However, the above statistical figures are shown in the most simplified form. The changes in the registered landlord-tenant relationships from 1972 to 1980 were more complicated. First, because of the "inheritance" of leased land by the original tenant's households, the landlord now has the *kuttagai* tenure with an increased number of tenants. In the three original cases, the number of tenants increased from 4 persons in 1972 to 19 persons in 1982. Second, of the 25 cases of the Peruvalanallur Hindu temples, 6 tenants returned the lands to the temples during the above period and new contracts have been made with some other tenants. Third, in 2 cases, the landlord asked his tenant to change the originally registered field to the other plots. Fourth, there were some cases where the registered tenants "sold" the right of cultivation to the other tenants without consulting the landlords, including the Hindu temple.

The Otti

Some characteristics of the *otti* system have already been pointed out in the introductory part to this chapter. Out of the 874 households in Peruvalanallur in 1979-80, 210 households or 24 percent were involved in *otti* tenancy of which 94 households were landowners (debtors) and 124 were tenants (creditors), although 8 households were involved in both. Like the case of the *kuttagai* tenancy, both landowners and tenants of Peruvalanallur had the *otti* transactions not only with their own villagers but also with other villagers (tables 20 and 21 and figs. 28, 29, and 30).

TABLE 20

OTTI TRANSACTIONS BY CASTE IN PERUVALANALLUR VILLAGE OF LALGUDI TALUK, TIRUCHIRAPPALLI DISTRICT, TAMIL NADU, INDIA

(1979-80)

I. Landowner Side

Castes	No. of House-Holds	Areas leased-out (in Acre) Wet	Dry	Total	"Credit" received (in Rs.)
Reddiar	31	56.76	1.09	57.85	458,050
Udaiyar	16	7.20	0.90	8.10	50,000
Gounder	9	3.22	8.75	11.97	26,600
Muslim	3	1.62	-	1.62	13,000
Muthuraja	5	2.46	-	2.46	17,500
Vannan	2	2.00	-	2.00	14,000
Pariyari	1	-	0.43	0.43	500
Pandaram	2	0.28	0.50	0.78	2,500
Pallan	18	6.30	0.85	7.15	47,170
Parayan	1	-	0.75	0.75	450
Catholic Pallan	1	0.28	-	0.28	2,500
Catholic Parayan	5	0.68	0.96	1.64	4,000
	94	80.80	14.23	95.03	636,270
Other Villagers	20	11.21	0.50	11.71	78,000
TOTAL	114	92.01	14.73	106.74	714,270

II. Tenant Side

Castes	No. of House-Holds	Areas leased-in (in Acre) Wet	Dry	Total	"Credit" given (in Rs.)
Reddiar	7	11.52	-	11.52	94,750
Udaiyar	22	14.56	-	14.56	111,250
Gounder	11	9.82	0.88	10.70	78,700
Muslim	2	1.88	-	1.88	23,000
Nadar	3	2.06	-	2.06	19,000
Achari	1	0.40	-	0.40	3,500
Muthuraja	2	2.99	-	2.99	16,500
Naidu	1	0.28	-	0.28	2,000
Mooppanar	1	0.22	-	0.22	2,000
Agampadiar	1	0.39	-	0.39	3,000
Pallan	47	26.87	3.64	30.51	191,370
Parayan	5	2.18	-	2.18	16,500
Catholic Pallan	8	5.42	-	5.42	44,000
Catholic Parayan	13	6.58	2.21	8.79	50,700
	124	85.17	6.73	91.90	656,270
Other Villagers	10	6.84	8.00	14.84	58,000
TOTAL	134	92.01	14.73	106.74	714,270

TABLE 21

<u>OTTI</u> TRANSACTIONS BETWEEN PERUVALANALLUR AND OTHER VILLAGERS OF LALGUDI TALUK, TIRUCHIRAPPALLI DISTRICT, TAMIL NADU, INDIA (1979-80)

I. Other villagers who leased-out to Peruvalanallur residents

Village Names of Landowners (Debtors)	Areas leased-out (in Acre)			"Credit" received
	Wet	Dry	Total	(in Rs.)
Thachankuruchi(#34)	0.40	-	0.40	3,500
Kumulur(#37)	1.22	0.50	1.72	9,300
Vellanur(#38)	0.75	-	0.75	4,000
Kanakkiliyanallur(#43)	0.37	-	0.37	3,000
Sirumayangudi(#62)	0.90	-	0.90	4,000
Poovalur(#64)	2.00	-	2.00	9,000
Mankkal(#65)	2.77	-	2.77	26,000
Sirudaiyur(#71)	0.62	-	0.62	3,000
Edaiyathumangalam(#81)	0.60	-	0.60	5,000
Perakambi(#86)	0.69	-	0.69	6,000
Siruganur(#91)	0.49	-	0.49	3,000
Garudamangalam(#104)	0.40	-	0.40	2,200
Total	11.21	0.50	11.71	78,000

II. Other villagers who leased-in from Peruvalanallur residents

Village Names of Tenants (Creditors)	Areas leased-in (in Acre)			"Credit" given
	Wet	Dry	Total	(in Rs.)
Reddimangudi(#35)	0.48	-	0.48	5,000
Kumulur(#37)	3.77	-	3.77	23,000
Sirumayangudi(#62)	1.37	-	1.37	11,000
Sirukalapur(#103)	1.22	-	1.22	14,000
Kottamulanur (Erode Dt.)	-	8.00	8.00	5,000
Total	6.84	8.00	14.84	58,000

Note: Each number in parentheses corresponds to the revenue village number listed in Table 1.

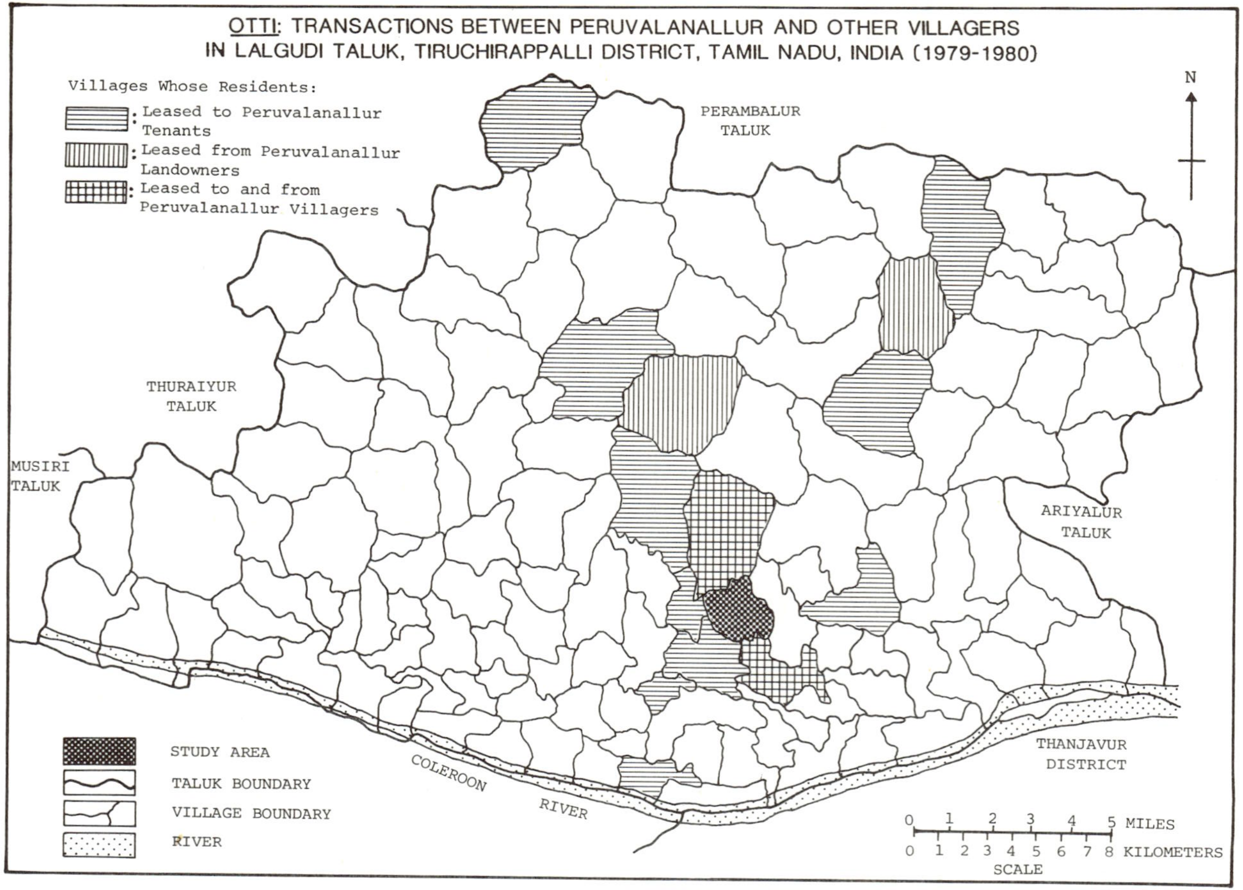
OTTI: TRANSACTIONS BETWEEN PERUVALANALLUR AND OTHER VILLAGERS
IN LALGUDI TALUK, TIRUCHIRAPPALLI DISTRICT, TAMIL NADU, INDIA (1979-1980)
Villages Whose Residents:
: Leased to Peruvalanallur Tenants
: Leased from Peruvalanallur Landowners
: Leased to and from Peruvalanallur Villagers
N
PERAMBALUR TALUK
THURAIYUR TALUK
MUSIRI TALUK
ARIYALUR TALUK
THANJAVUR DISTRICT
COLEROON
RIVER
STUDY AREA
TALUK BOUNDARY
VILLAGE BOUNDARY
RIVER
0 1 2 3 4 5 MILES
0 1 2 3 4 5 6 7 8 KILOMETERS
SCALE

Fig. 28

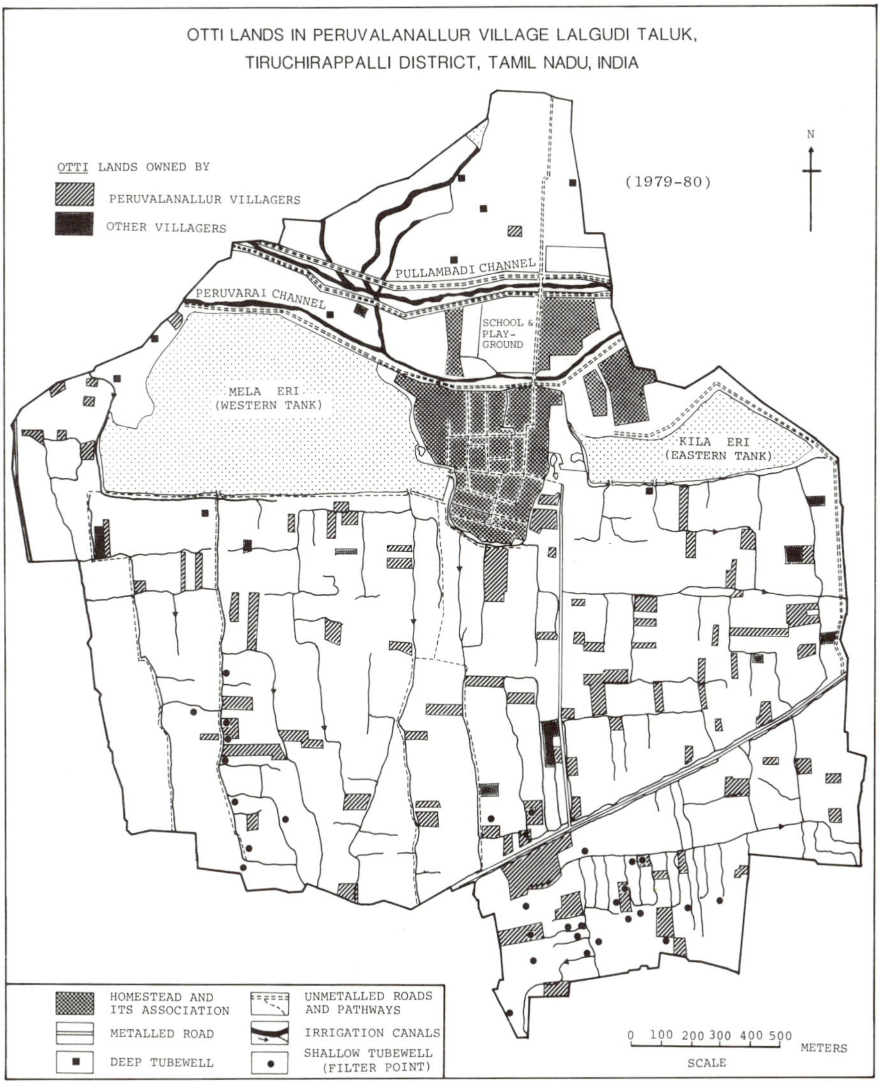

OTTI LANDS IN PERUVALANALLUR VILLAGE LALGUDI TALUK,
TIRUCHIRAPPALLI DISTRICT, TAMIL NADU, INDIA
OTTI LANDS OWNED BY
PERUVALANALLUR VILLAGERS
OTHER VILLAGERS
(1979-80)
N
PULLAMBADI CHANNEL
PERUVARAI CHANNEL
SCHOOL & PLAY-GROUND
MELA ERI
(WESTERN TANK)
KILA ERI
(EASTERN TANK)
HOMESTEAD AND ITS ASSOCIATION
METALLED ROAD
DEEP TUBEWELL
UNMETALLED ROADS AND PATHWAYS
IRRIGATION CANALS
SHALLOW TUBEWELL (FILTER POINT)
0 100 200 300 400 500
METERS
SCALE

Fig. 29

OTTI: AREAS LEASED-OUT AND -IN BY PERUVALANALLUR VILLAGERS IN LALGUDI TALUK, TIRUCHIRAPPALLI DISTRICT, TAMIL NADU, INDIA (1979-80)

(Unit: in Acre)

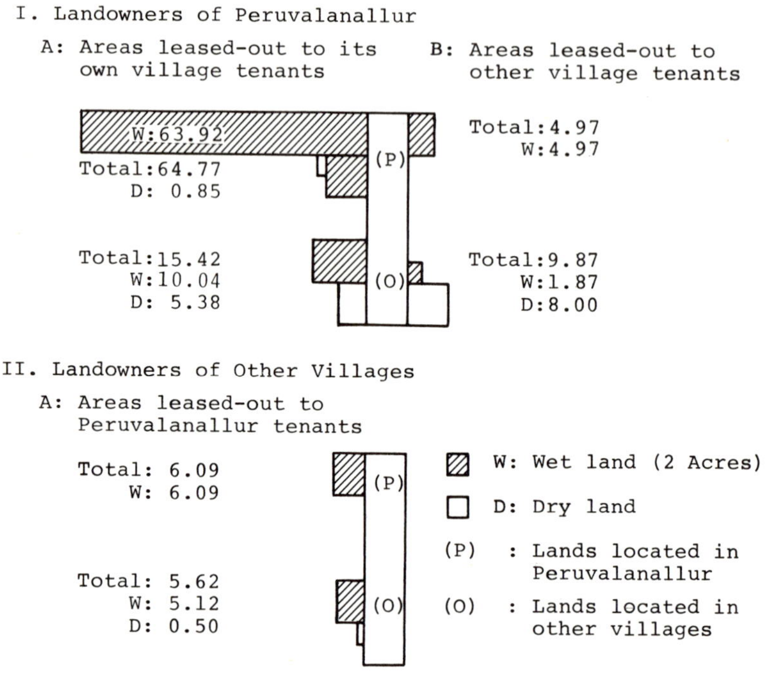

Fig. 30

The 94 landowner households in Peruvalanallur together received Rs. 636,270 of the cash "credits" in exchange for the leasing of 95.03 acres (wet: 80.80 acres; dry: 14.23 acres) under the *otti* which were located not only in Peruvalanallur (69.74 acres) but also in its neighboring villages (25.29 acres). Of these same 95.03 acres of *otti* lands leased by the Peruvalanallur villagers, the tenants of the same village cultivated 80.19 acres (wet: 73.96 acres; dry: 6.23 acres), and the other village tenants cultivated the remaining 14.84 acres (wet: 6.84 acres; dry: 8.00 acres). The lands involved in these transactions were predominantly wet land except for the land owned by residents of Peruvalanallur leased out to residents of other villages in which dry land slightly exceeded wet land.

The tenants of Peruvalanallur had also *otti* tenure relationships with landowners of other villages and towns for 11.71 acres (tables 20 and 21 and figs. 28 and 30). These lands were located not only in Peruvalanallur (6.09 acres of wet land) but also in its adjacent villages, 5.62 acres (wet: 5.12 acres; dry: 0.50 acre) (fig. 30). More specifically, the *otti* lands leased to the Peruvalanallur tenants by the landowners in some distant villages such as Perakambi (no. 86), Garudamangalam (no. 104), Siruganur (no. 91), and Kanakkiliyanallur (no. 43) listed in table 21 were entirely located within Peruvalanallur. Above all, as far as the Peruvalanallur tenants were concerned, they cultivated the lands which were located a fairly short distance from their own village (mostly within 5 kms, but at most about 8 kms). However, this is not applicable for the *otti* tenants in the dry villages: they cultivated the *otti* lands in Peruvalanallur, which were located a long distance from their respective villages (about 18 kms at most). This fact is certainly related to the general agro-economic situations in the dry villages. There were fewer chances for profitable agricultural investment in their own villages, and the *otti* lands themselves were scarcely available in and around their respective villages.

To summarize, the tenants of Peruvalanallur together leased 91.90 acres (wet: 85.17 acres; dry: 6.73 acres) of *otti* lands not only from their own villagers (80.19 acres) but also from the other village and town dwellers, mostly neighboring villagers (11.71 acres), and paid Rs. 656,270 for the right to cultivate the areas involved.

At this stage, some important aspects should be pointed out. First, out of the total area involved in the *otti* tenure in the studied area in 1979-80, the ratio for wet land was extremely high compared with that for dry (tables 14 and 20 and figs. 29 and 30). Even when comparing the *otti* and *kuttagai,* it was much higher in the former (86 percent) than in the latter (75 percent). This is certainly related to the fact that the wet lands generally provide higher productivity under relatively stable physiograhic conditions for cultivation (although there is a great variation within them). Second, unlike the case of the *kuttagai* tenancy, out of the total *otti* areas leased by the Peruvalanallur landowners, the other village tenants had a very small share compared with that of the Peruvalanallur tenants,[20] although the total areas of the *kuttagai* and *otti* lands differed greatly (205.24 acres of the *kuttagai*; 95.03 acres for the *otti).* Third, the spatial transactions of the *otti* between Peruvalanallur and the other villagers were very limited in comparison with those of the *kuttagai* (figs. 22 and 28).

As already mentioned, the tenants' profit from cultivating the *otti* land is regarded as his "interest" on the amount of his cash deposits ("credits") give to the landowner. Thus, it is beneficial for the tenant to pay as little money as possible per unit of area in order to get the equivalent "interest" on the "loan" as high as possible, and vice versa for the landowner. There was serious bargaining for the settlement of the amount of "credit" between the landowner and the would-be tenants before the final decision was made. The actual amount of "credits" given by the tenants varied greatly from Rs. 500 to Rs. 11,475 per acre, with Rs. 6,692 being the average. More specifically, the average amount of "credit" per acre for the wet land was Rs. 7,629 with a range of Rs. 2,500 - Rs. 11,475, while for the dry land it was Rs. 835 with a range of Rs. 500 - Rs. 2,000.

There are some important factors responsible for these variations. For the wet land the amount of "credit" varied widely, depending upon whether the land was a "single" or "double" cropping area, and for the dry land whether it was equipped with irrigation facilities or not. Thus, the amount of "credit" for the *otti* land was settled primarily on the quality of the land involved. It should be noted that data shown in the above tables and figures include all the cases involving the current *otti* tenure by the Peruvalanallur villagers observed during the year of 1979-80, regardless of when the individual contracts were made, and that the

---

20. Of the total areas of both the *kuttagai* and *otti* lands leased from the Peruvalanallur landowners, the tenants of the other villages cultivated 73.08 percent or 149.98 acres (wet: 82.13 acres; dry: 67.85 acres) for the *kuttagai* but only 15.09 percent or 14.34 acres (wet: 6.16 acres; dry: 8.00 acres) for the *otti.*

amounts of the "credit" were based on the actual payments and/or receipts at the time of each contract. An examination of the amounts of the "credit" for each individual case shows that they were based mostly on 55-65 percent of the market price of the land involved at the time of the contracts, although the market prices themselves have been increasing at a high rate in recent years, especially in the latter half of the 1970s.[21] As the terms for most of the *otti* contracts were for at least three years, the amount of "credit" for the same quality of land per unit of area was assumed to be the highest on the latest contracts. Thus, the time of the contract in the *otti* tenure is another factor responsible for the variation in the amount of the "credit" which appears in our statistics.

Let us examine the tenant's profit from cultivating the *otti* land which can be expressed in terms of the "annual interest" on the amount of his cash deposit given to the landowner. Among the varied qualities of land involved in the *otti,* we take into account here three typical types of land: (1) dry land, (2) "single" cropping wet land, and (3) "double" cropping wet land. In chapter III, we have already pointed out some basic characteristics of land-use and crop associations in relation to the physiographic conditions and tried to figure out yields, costs, and benefits of the individual crops (see table 6). On the other hand, the average amount of the "credit" per acre for dry, "single-wet," and "double-wet" land in 1979-80 was Rs. 835, Rs. 3,000, and Rs. 7,000 respectively. Based on these data, we can get the "net benefit" in terms of the annual rates of "interest" on the amount of "credit" for the different types of crop associations. It should be noted that all the available types of crop associations in Peruvalanallur were also observed in the *otti* lands, because the *otti* lands in the studied village were distributed almost evenly regardless of the different physiographic conditions of cultivation (fig. 29). The obtained figures in table 22 reveal considerably high rates of annual interest with a range of 30-55 percent. Although these rates cannot be claimed to be perfectly accurate,[22] they correspond well to the villagers' general understanding that the *otti* tenants can safely get back one-third of the amount of "credit" given to their landowners in a year. Realizing that a favorable annual interest

---

21. For example, in March of 1980, the land value of good wet land was Rs. 18,000 per acre in the village, and, after 18 months, the value of the same quality of land increased to Rs. 30,000 per acre.

22. In fact, there is an impression that the annual rate of "interest" or "net benefit" seemed to be a little higher for double cropping wet land used for the sugar cane and gram.

TABLE 22

TENANT'S PROFITS IN THE OTTI TENANCY BASED ON THE DIFFERENT TYPES OF CROP ASSOCIATIONS IN PERUVALANALLUR OF LALGUDI TALUK, TIRUCHIRAPPALLI DISTRICT, TAMIL NADU, INDIA

(1979-80)

Types of Crop Associations	(a) Average "Benefit" per Acre (in Rs.)	(b) Average "Credit" per Acre (in Rs.)	(c) "Net Benefit" [Rates of Interest in % (a) ÷ (b) x 100]
1. Dry Land		835	
Single Cropping Land	250		29.94
2. Single Cropping Wet Land		3,000	
(a) Samba Paddy	960		32.00
(b) Samba - Gram	1,460		48.66
3. Double Croppping Wet Land		7,000	
(a) Kuruvai - Thaladi	2,280		32.57
(b) Kuruvai - Thaladi - Gram	2,780		39.71
(c) Sugarcane	3,325		47.50
(d) Sugarcane - Gram	3,825		54.64

Notes: 1. "Benefit" under (a) column refers to the annual "net benefit" obtained by the ordinary landowners (cf. Table 6).

2. Under the otti tenancy, the tenant still has to pay the rent as a part of the costs. However, the rent is paid in terms of cash "credit" (deposit) to his landowner. The amount of "credit" per unit of area varies greatly depending upon the quality of land as shown under (b) column.

for rural people deposited in the authorized banks was 10-15 percent, it is true in a sense that the tenants' involvement in the *otti* tenure can be regarded as a "positive investment."

Like the case of the *kuttagai* tenure, one of the important issues of the *otti* tenure is that whether or not there are any relationships between the involved households (as landowners and as tenants) and the particular socio-economic groups in the rural community. In this respect, we will examine the involved households in the *otti* in relation to caste and size of landholding.

With respect to the caste structure of landowners and tenants involved in *otti* transactions, *Reddiar* predominate among landowners with 61 percent of the area leased and 72 percent of the amount received in rupees, while among tenants *Pallan* with 34 percent of the area and 29 percent of the payments and *Udaiyar* with 16 percent of area and 17 percent of payments are the leading castes (based on figures in table 20). Figure 31 shows in detail the transactions involved among each of the castes. As would be expected, transactions between *Reddiar* as landowners (debtors) and *H. Pallan* and *Udaiyar* as tenants (creditors) are the most numerous, although transactions between *Reddiar* and *Reddiar,* though involving fewer transactions, involve relatively large amounts of money.

Table 23 and figures 32 and 33 show the distribution of the involved households, areas, and "credits" in the *otti* tenancy by the size of the landholding in the studied village. The above tables and figures reveal certain characteristics. Out of the 94 total landowners' households in the *otti* tenancy in Peruvalanallur, 50 percent (or 47 households) fall into two categories (under 1 acre and 1-2 acres), and they occupied 24 percent (or 22.60 acres) of the total area leased and 20 percent (or Rs. 128,020) of the total amount of money deposits received in the studied village. On the other hand, only 24 percent (or 10 households) in the largest category (15 acres and above) occupied 33 percent (or 30.73 acres) of the area and 40 percent (or Rs. 252,200) of the "credit" of the respective totals. In other words, there was a great difference in the extent of area leased and the amount of money-deposit received as "credit" per household between the marginal groups (under 2 acres) and the greatest landholding groups (more than 15 acres); that is, the former group leased 0.48 acres and received Rs. 2,724 per household, and the latter groups leased 3.07 acres and received Rs. 25,222 per household.

By definition, landless households can only participate in the *otti* tenure on the tenant side. However, 3 households which

RELATIONSHIPS AMONG LANDOWNERS AND TENANTS UNDER OTTI BY CASTE IN PERUVALANALLUR VILLAGE, LALGUDI TALUK, TIRUCHIRAPPALLI DISTRICT, TAMIL NADU, INDIA (1979-1980)

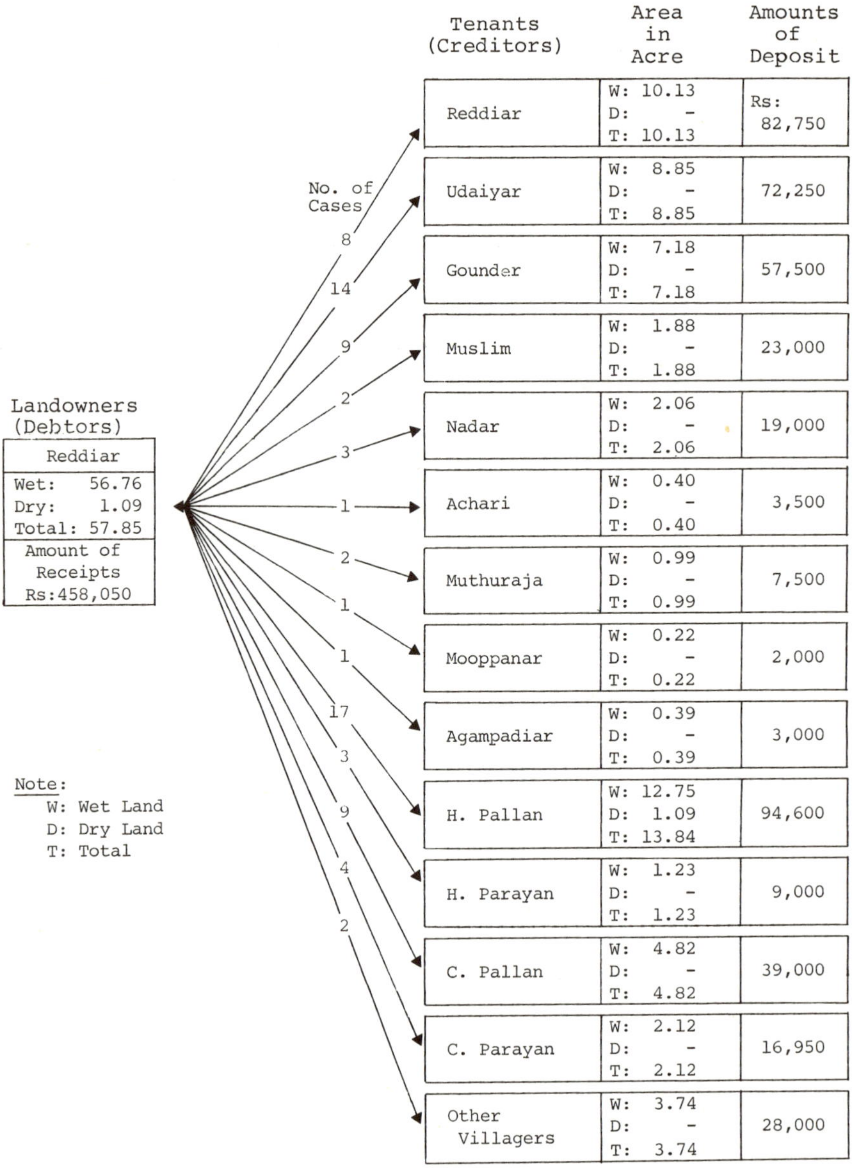

Fig. 31

OTTI

Landowners (Debtors): Udaiyar

Udaiyar	
Wet:	7.20
Dry:	0.90
Total:	8.10
Amount of Receipts	Rs:50,000

Nos. of Cases	Tenants (Creditors)	Area in Acre	Amounts of Deposit
6	Udaiyar	W: 3.13 D: - T: 3.13	Rs: 20,000
6	H. Pallan	W: 1.92 D: 0.90 T: 2.82	14,000
3	C. Parayan	W: 1.32 D: - T: 1.32	9,000
2	Other Villagers	W: 0.83 D: - T: 0.83	7,000

Gounder	
Wet:	3.22
Dry:	8.75
Total:	11.97
Amount of Receipts	Rs:26,600

Nos. of Cases	Tenants (Creditors)	Area in Acre	Amounts of Deposit
2	Gounder	W: 0.40 D: 0.45 T: 0.85	3,500
6	H. Pallan	W: 2.34 D: 0.30 T: 2.64	13,100
3	Other Villagers	W: 0.48 D: 8.00 T: 8.48	10,000

Muslim	
Wet:	1.62
Dry:	-
Total:	1.62
Amount of Receipts	Rs:13,000

Nos. of Cases	Tenants (Creditors)	Area in Acre	Amounts of Deposit
1	Udaiyar	W: 0.70 D: - T: 0.70	7,000
2	Other Villagers	W: 0.92 D: - T: 0.92	6,000

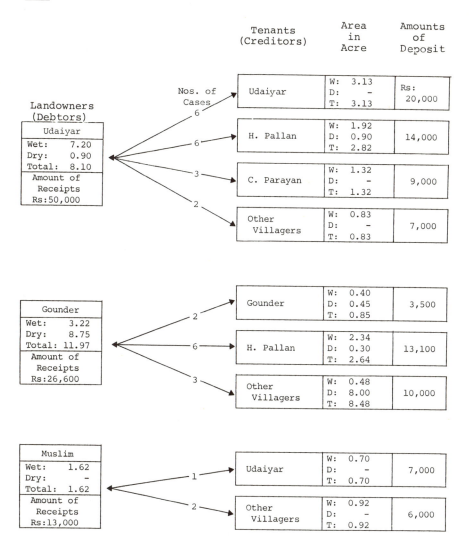

OTTI

Landowners (Debtors)	Nos. of Cases	Tenants (Creditors)	Area in Acre	Amounts of Deposit
Muthuraja Wet: 2.46 Dry: - Total: 2.46 Amount of Receipts Rs:17,500	1	Udaiyar	W: 0.33 D: - T: 0.33	Rs: 2,500
	3	H. Pallan	W: 1.26 D: - T: 1.26	8,000
	1	Other Villagers	W: 0.87 D: - T: 0.87	7,000
Vannan Wet: 2.00 Dry: - Total: 2.00 Amount of Receipts Rs:14,000	2	H. Pallan	W: 2.00 D: - T: 2.00	14,000
Pariyari Wet: - Dry: 0.43 Total: 0.43 Amount of Receipts Rs: 500	1	Gounder	W: - D: 0.43 T: 0.43	500
Pandaram Wet: 0.28 Dry: 0.50 Total: 0.78 Amount of Receipts Rs: 2,500	2	H. Pallan	W: 0.28 D: 0.50 T: 0.78	2,500

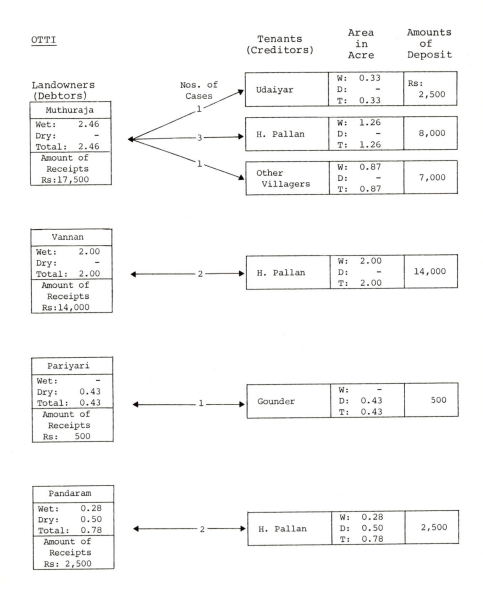

OTTI

Landowners (Debtors)	Wet	Dry	Total	Amount of Receipts	Nos. of Cases	Tenants (Creditors)	Area in Acre W	Area in Acre D	Area in Acre T	Amounts of Deposit
H. Pallan	6.30	0.85	7.15	Rs:47,170	3	Udaiyar	0.85	-	0.85	5,000
					2	Gounder	1.09	-	1.09	9,000
					13	H. Pallan	3.12	0.85	3.97	23,170
					2	C. Parayan	1.24	-	1.24	10,000
H. Parayan	-	0.75	0.75	Rs: 450	1	C. Parayan	-	0.75	0.75	450
C. Pallan	0.28	-	0.28	Rs: 2,500	1	H. Parayan	0.28	-	0.28	2,500
C. Parayan	0.68	0.96	1.64	Rs: 4,000	1	Naidu	0.28	-	0.28	2,000
					4	C. Parayan	0.40	0.96	1.36	2,000

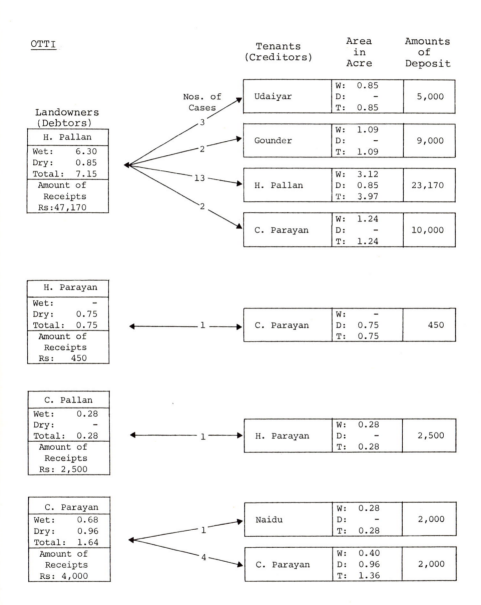

OTTI

Landowners (Debtors)

Other Villagers	
Wet:	11.21
Dry:	0.50
Total:	11.71
Amount of Receipts	Rs:78,000

Nos. of Cases	Tenants (Creditors)	Area in Acre	Amounts of Deposit
2	Reddiar	W: 1.39 D: - T: 1.39	Rs: 12,000
2	Udaiyar	W: 0.70 D: - T: 0.70	4,500
2	Gounder	W: 1.15 D: - T: 1.15	8,200
1	Muthuraja	W: 2.00 D: - T: 2.00	9,000
8	H. Pallan	W: 3.20 D: - T: 3.20	22,000
1	H. Parayan	W: 0.67 D: - T: 0.67	5,000
1	C. Pallan	W: 0.60 D: - T: 0.60	5,000
3	C. Parayan	W: 1.50 D: 0.50 T: 2.00	12,300

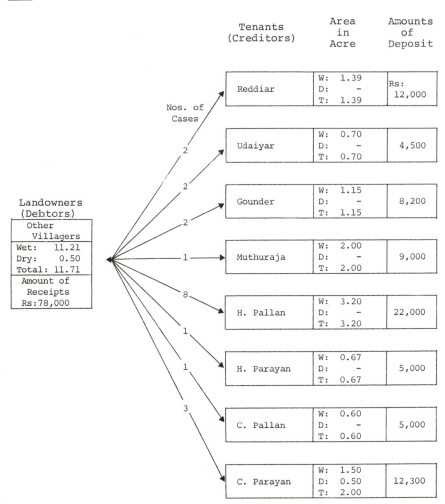

TABLE 23

OTTI: THE INVOLVED HOUSEHOLDS, AREAS, AND "CREDITS" BY SIZE OF LANDHOLDING IN PERUVALANALLUR VILLAGE OF LALGUDI TALUK, TIRUCHIRAPPALLI DISTRICT, TAMIL NADU, INDIA (1979-80)

I. Landowners ("Debtors") Side

Categories by Size of Land-holding	No. of House-holds	Areas-leased-out			"Credit" received for		
		Wet Land (in Acre)	Dry Land (in Acre)	Total (in Acre)	Wet Land (in Rs.)	Dry Land (in Rs.)	Total (in Rs.)
C-1: Landless	3	0.96	-	0.96	6,200	-	6,200
C-2: under 1 acre	32	10.03	3.36	13.39	69,620	3,900	73,520
C-3: 1-2 acres	15	7.43	1.78	9.21	52,600	1,900	54,500
C-4: 2-3 acres	8	4.46	-	4.46	31,750	-	31,750
C-5: 3-5 acres	6	7.55	4.00	11.55	54,250	3,000	57,250
C-6: 5-7 acres	8	5.53	-	5.53	45,000	-	45,000
C-7: 7-10 acres	5	2.91	1.09	4.00	21,550	1,300	22,850
C-8: 10-15 acres	7	11.20	4.00	15.20	91,000	2,000	93,000
C-9: 15 & above	10	30.73	-	30.73	252,200	-	252,200
Total	94	80.80	14.23	95.03	624,170	12,100	636,270

II. Tenants ("Creditors") Side

Categories by Size of Land-holding	No. of House-holds	Areas leased-in			"Credit" given for		
		Wet Land (in Acre)	Dry Land (in Acre)	Total (in Acre)	Wet Land (in Rs.)	Dry Land (in Rs.)	Total (in Rs.)
C-1: Landless	61	34.41	0.76	35.17	256,600	1,000	257,600
C-2: under 1 acre	33	23.72	3.66	27.38	175,450	3,700	179,150
C-3: 1-2 acres	19	7.70	1.96	9.66	67,100	2,000	69,100
C-4: 2-3 acres	5	2.88	0.35	3.23	20,770	700	21,470
C-5: 3-5 acres	4	2.93	-	2.93	25,000	-	25,000
C-6: 5-7 acres	4	2.50	-	2.50	17,200	-	17,200
C-7: 7-10 acres	3	2.85	-	2.85	23,500	-	23,500
C-8: 10-15 acres	4	4.83	-	4.83	37,250	-	37,250
C-9: 15 & above	1	3.35	-	3.35	26,000	-	26,000
Total	124	85.17	6.73	91.90	648,870	7,400	656,270

OTTI: PERCENTAGE DISTRIBUTION OF HOUSEHOLDS AND AREAS INVOLVED IN OTTI BY SIZE OF LANDHOLDING IN PERUVALANALLUR (1979-80)

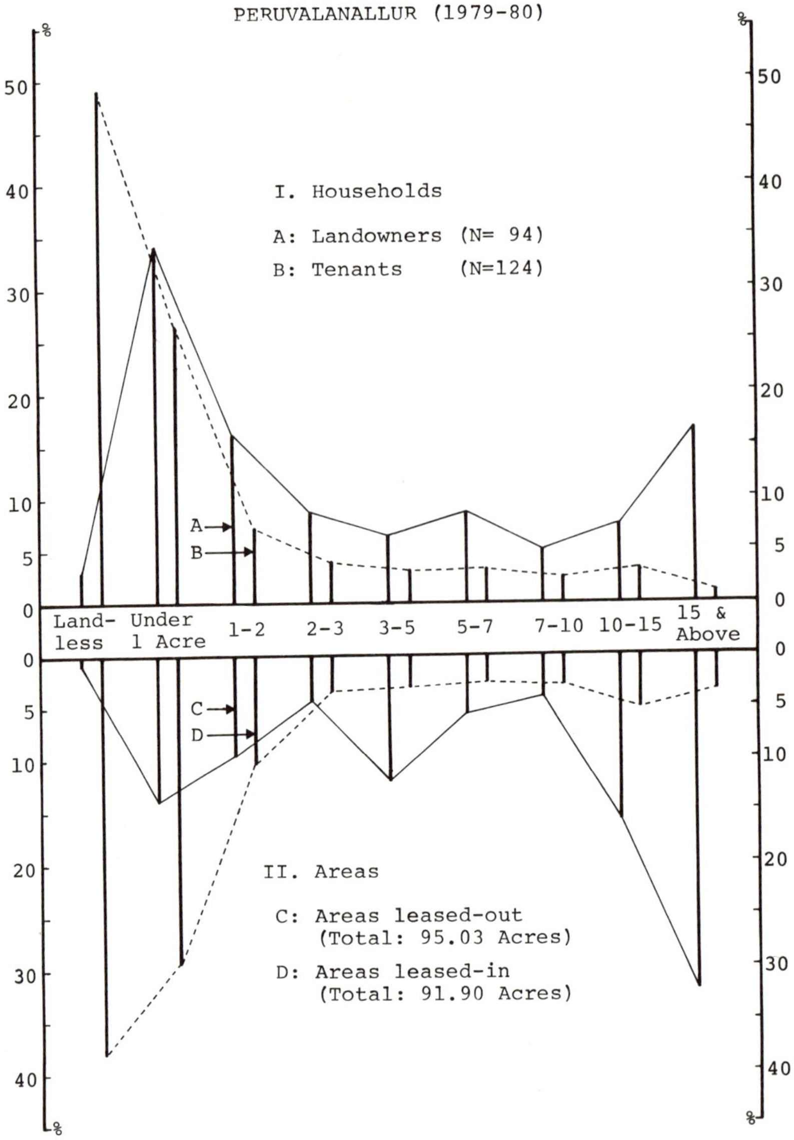

Fig. 32

OTTI: PERCENTAGE DISTRIBUTION OF AREAS AND AMOUNTS OF MONEY INVOLVED IN OTTI BY SIZE OF LAND-HOLDING IN PERUVALANALLUR (1979-80)

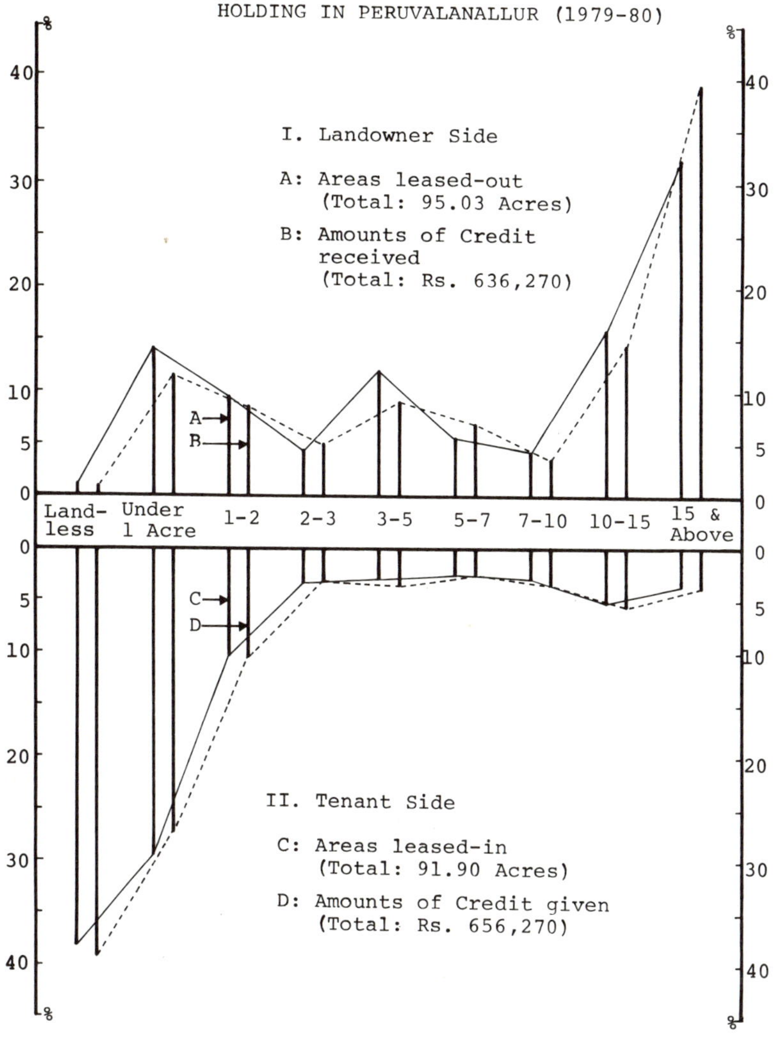

Fig. 33

appeared in the tables and figures got *kuttagai* lands (which belonged to the Hindu temples), and then they leased the lands to others under the *otti* tenancy.

Although the *otti* landowners' ultimate motivation is mainly to secure a large amount of cash money promptly, there were various reasons among the involved households, as indicated below, why they decided to pledge their lands under the *otti* at that time:

1. For investment in new business such as agents for fertilizer sales, shops for fertilizers and agricultural medicines, work shops for agricultural implements and machines, tractors for plows and carraiges, etc.
2. For higher education and technical training, job promotion and/or seeking a job itself, etc.
3. For expenses for marriage (especially on the bride's side)
4. For general family maintenance
5. Due to the lack of working members for agriculture within the families.

Of course, there were many other reasons or motivations for the *otti* landowners. Moreover, there were usually not one but two or more reasons for each case of the *otti* transactions depending on the socio-economic conditions of the individual households involved.

Most of the *otti* relationships involve one tenant and one landowner (table 24), but of the 94 landowners 16 had 2 tenants, 5 had 3 tenants, 1 had 5 tenants, and 2 had 11 tenants apiece. Of the 124 tenants 20 had relations with 2 landowners, 3 with 3 landowners, and 2 with 4 landowners.

Compared with the *kuttagai* contracts (fig. 27), many of which have run for 20, 30 or 40 years, the *otti* contracts are of relatively short duration (fig. 34). The *kuttagai* contracts imply a continuing landowner-tenant relationship in which the right of the tenants are imbedded in legislation, whereas the *otti* contracts are short-term highterest borrowing by landowners from those willing to work the land.

TABLE 24

<u>OTTI</u>: NUMBER OF COUNTERPARTS FOR EACH LANDOWNER AND TENANT IN PERUVALANALLUR VILLAGE OF LALGUDI TALUK, TIRUCHIRAPPALLI DISTRICT, TAMIL NADU, INDIA (1979-80)

I. Landowner (Debtor) Side		II. Tenant (Creditor) Side	
Number of Counterparts	Number of Cases (Landowners' Households)	Number of Counterparts	Numberof Cases (Tenants' Households)
1 tenant	70 (74.47%)	1 landowner	99 (79.84%)
2 tenants	16 (17.02)	2 landowners	20 (16.13)
3 tenants	5 (5.32)	3 landowners	3 (2.42)
5 tenants	1 (1.06)	4 landowners	2 (1.61)
11 tenants	2 (2.13)		
	94 (100.00%)		124 (100.00%)

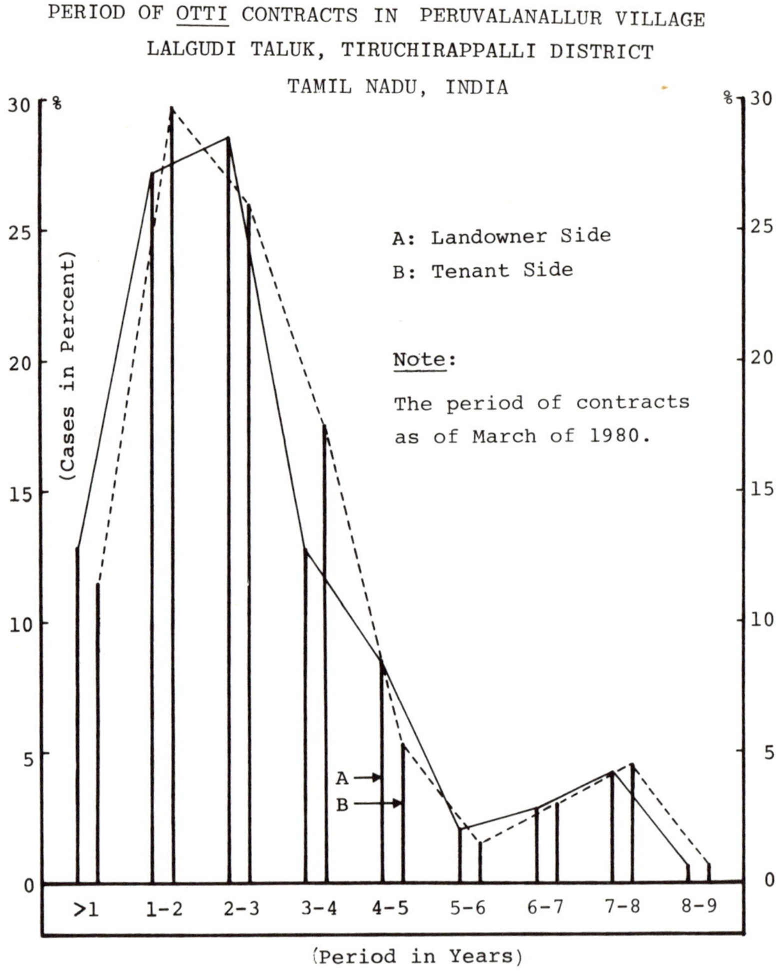

PERIOD OF OTTI CONTRACTS IN PERUVALANALLUR VILLAGE
LALGUDI TALUK, TIRUCHIRAPPALLI DISTRICT
TAMIL NADU, INDIA
A: Landowner Side
B: Tenant Side
Note:
The period of contracts as of March of 1980.
(Cases in Percent)
%
0
5
10
15
20
25
30
A
B
>1
1-2
2-3
3-4
4-5
5-6
6-7
7-8
8-9
(Period in Years)

Fig. 34

CHAPTER VII
THE CHANGING VILLAGE

In this chapter some concluding remarks will be made about land use, land ownership, land tenure, the socio-economic structure, and agricultural development in Peruvalanallur village. This study has been concerned fundamentally with rural communities and agricultural systems in South India, with specific reference to the variations, both geographical and functional, found in socio-economic activities. More specifically, this study had the following objectives: first, to clarify the structural and spatial patterns of a rural community with special reference to its agricultural activities and to measure the degree of association and integration among villagers whose socio-economic background seem to vary to a great extent; second, to investigate recent changes in the socio-economic aspects of the community; and by extension, third, to identify some important elements in its modernization.

In order to meet these objectives, the research has focused on the following four topics: (1) land uses and their associations, (2) landownership, (3) occupational specialization and labor organization, and (4) land tenure. The analytical methods adopted included conventional ecological and spatial concepts appropriate for the study of important aspects of a rural community and its development.

This study dealt primarily with a selected community, Peruvalanallur village of Lalgudi Taluk of Tiruchirappalli (Tiruchy) District, Tamil Nadu State in the Republic of India. The studied village, lying on the left bank of the Coleroon River (a distributary of the Cauvery River), is located some 26 km northeast of Tiruchy (the district capital) and some 6 km east-northeast of Lalgudi, where the *taluk* head office is located. In terms of agricultural activities, Lalgudi Taluk can be classified largely into a wet zone and a dry zone. The wet zone corresponds to the alluvial plains distributed narrowly along the Coleroon River. The agricultural activities in this zone have been dependent upon the development of irrigation. There are in the wet zone several irrigation channels, which are connected with the Upper Anicut of the Cauvery River. The wet zone can be further divided into two physiographic sub-zones; that is, a "lower wet" zone and an "upper wet" zone. The availability of irrigation water is more stable in the former than in the latter. The studied village belongs to the "upper wet" zone.

The vast area of the dry zone extends to the north of the upper wet zone in the *taluk* and is highly dissected along the non-perennial jungle streams (*varis*). In the dry zone the water for agricultural use is mostly dependent upon the monsoon rains during October-December. However, in many villages of the dry zone in the *taluk,* there are some "wet land" areas being irrigated from different sources: *eris* (tanks), traditional wells, and ground water in which irrigation techniques themselves have been changing recently. It should be noted that these model physiograhic zones can be found in much of the Cauvery River basin, although individual areas may vary to a great extent depending upon the regional units concerned.

Peruvalanallur has a territory of 1,335.70 acres. There are two big tanks for irrigation and other uses: the *mela eri* (western tank) and the *kila eri* (eastern tank), extending to the west and east of the major residential areas of the village. Being connected with the Peruvalai channel, which is controlled at the Upper Anicut by the Public Works Department, the tanks provide water to the southern fields of the village through well-developed field canal networks. Thus, most of the wet lands in the village extend southward of the two tanks. "Dry lands" on higher elevations are found mostly in the northern section of the village; some small irrigated pockets are located within the dry lands.

Residential Patterns

The residential patterns in Peruvalanallur seem to reflect clearly the segregation of the caste systems prevailing in rural Indian communities. In 1980, the residents of the village consisted of 33 caste groups and 874 households (3,496 persons). Of these caste groups, eight (*Reddiars*, *Udaiyars*, *Gounders*, Muslims, *Pallans*, *Parayans*, Catholic *Pallans*, and Catholic *Parayans*) had a considerable number of households, a sizeable population, and distinct individual residential areas, which originally centered around the *Reddiars*, the dominant caste in the village. Some of the other caste groups resided together on the peripheries of the major caste groups. More specifically, they desire to have in their neighborhood closely-ranked castes, and preferably the higher ones, rather than the lower.

Some other members of minor castes live within the residential area of the major caste groups, without forming their own residential areas. There are few communal troubles between the "intruder" of the minor caste and its immediate neighbors (and the community) of the major caste, if and when the caste rankings of the minor castes were higher than those of the major castes. Such a case is seen between the *Brahmans*, as a minor caste, and the *Reddiars*, as a major caste in the village. Indeed, the *Brahmans* reside in the core of the *Reddiar* community. However, if these caste rankings were reversed, the major caste groups would put their communal pressure one way or another not only on the "invader," but also on the members of the major castes who had associated with the "invader." A serious case for concern in this community occurred during the period of this field work; a *Harijan* family occupied one of the *Reddiar*'s residences within the *Reddiar* community.

Expansion of the Residential Area and Its Implications

In relation to the development of residential areas in Peruvalanallur, valuable historical documents are available: the *First Settlement Register* (1864), the *First Re-settlement Register* (1898), and the *Second Re-settlement Register* (1927), as well as their respective cadastral maps, which were mostly reconstructed and modified by the author. During a long period from 1864 to 1927, there was no substantial expansion of the residential area, although during 1898-1927 very minor changes occurred in its northeastern periphery, none of which changed the areal extent. It is clear from the data that the core of the residential area in 1864 corresponded to that of the present *Reddiar* community. It should be noted that in 1864 the *Reddiars* as a caste group occupied over 90 percent of the privately-owned land in the village[1] and that a few of the *Reddiar* landlords were assigned to supervise *poramboke* (common land) such as tanks, public roads, river, etc. Although we do not know when people started to settle in this village, the above evidence suggests that the *Reddiars* were socio-economically the dominant caste group from an early stage of the village settlement.

---

1. Since the Settlment Registers include the landowners in other villages who owned lands in Peruvalanallur, this figure would be less if we account for lands owned only by the residents of the village.

During 1927-1980, there was a great expansion of the residential areas in the village, the expanded area almost equalling the entire residential area of 1927. Although the expanded area during the period can be observed in many places in the village, the major areas correspond to the present communities of "Muslim Street," the "North Hindu *Pallan,"* and "East *Harijan"* (within which Hindu *Pallan,* Hindu *Parayan,* and Catholic *Parayan* castes have sizeable residential areas respectively).

The villagers create new residences in three ways: (1) by purchasing open fields (mostly higher land), (2) by renting banks of irrigation channels and some open spaces along the public roads having contracts with the Public Works Department (PWD), and (3) by creating a new "colony" under the local government program with the cooperation of some big landlords and the community itself.

The first two ways generally result in a piecemeal expansion of the residential areas, whereas the last method provides an organized expansion scheduled for completion in a limited time. The first type can be observed in the "Muslim Street" to the west, a new Muslim area across the Peruvalai channel to the north, and most of the "mixed caste" areas of Peruvalanallur. The second type is well exemplified in the "service area" to the immediate south of the Peruvalai channel, near the central portion.

The third type is represented by the "North Hindu *Pallan"* area and the "East Hindu *Pallan"* area (which is part of the "East *Harijans*"). The planning of the "North Hindu *Pallan* Colony" began following a fire that occurred in the "South *Pallan"* community in 1920. Land for the "new colony" was acquired from some big *Reddiar* landlords by the government and allocated to the victims of the fire free of cost. A long-term housing loan was provided to each family under the newly-formed Housing Co-operatives. However, the intra-village migration of the victims from the "South" to the "North" continued for about ten years following the fire.

The planning of the "East Hindu *Pallan* Colony" also was initiated as a result of another big fire that took place in the "South Hindu *Pallan"* community in 1963. The process of planning and execution was similar to the case of the "North Hindu *Pallan* Colony," but it took only a few years to shape the "new colony" (for only the well-to-do families migrated from the "South" to "East").

Besides the above types, we can find many cases of "illegal occupation" of *poramboke* (common land) and government land by immigrants, most of whom are from dry villages. This type can probably be applied in the case of the Hindu *Parayan* and the

Catholic *Parayan* castes who have lived on the outskirts (immediately north of *kila eri*) of the major residential area even before 1927[2] and who, by 1980, together occupied almost half of the "East *Harijan*" residential area.

The expansion of the residential area in a given rural community generally is related to its population growth, normally the outcome of both natural growth within the community and migration from outside. However, we should not underestimate the number of immigrants to this particular village. The results of our interviews support an assumption that there was a great number of immigrants from outside the village during the period 1927-1980 and that they were mostly landless agricultural laborers regardless of their castes.

"What were the reasons for this migration to this particular village?" In answer to this question, it should be noted that there had been great ecological changes in agricultural lands in the village during the period 1927-1980: a great conversion from "single-cropping" to "double-cropping" wet lands supported by the development of irrigation systems. This change certainly has created more jobs in agricultural activities in the village.

Inter-village Landholdings

In Peruvalanallur in 1980, agricultural land, including uncultivated areas, accounted for 945.70 acres, or 71 percent of the village territory. It is confirmed that most of the agricultural land in Peruvalanallur was owned by the village residents and Hindu temples (and other public bodies) but that some was owned by outsiders (including Hindu temples). It is also true that Peruvalanallur villagers owned much land in many other villages, although the lands were distributed mostly in the neighboring villages within the same *taluk*.

It should be noted that the Peruvalanallur villagers together owned much more land in other villages than in their own village and that in these other villages they owned mainly dry land (2.5 times as much as wet land). These facts themselves contain some important problems related to the socio-economic activities of the villagers.

When someone holds land in a neighboring village, the owner regards the land as a part of his immediate farm unit, since the owner's land in the other village is within a manageable distance of his residence. This applies to most of the cases between the

---

2. In this kind of case, the village authority *(Reddiar)* usually did not report to the local government unless the villagers were disturbed. In those days, as today, the immigrants were poor agricultural laborers who worked mostly for the *Reddiar*s,

studied village and its neighboring villages.

There is another type of landholding, however, whereby the Peruvalanallur villagers own land in more distant villages. For this type of landholding the locations of the lands involved correspond mostly to the "mother villages" of the owner's family members (and their ancestors). This is also true for the outside villagers (and town dwellers) who have land in Peruvalanallur. This dispersion of holdings thus is mainly the result of inheritance and of inter-village marriages.

Migration of a family unit to the studied area is also responsible for some landholding in distant villages, since the migrants have usually retained their land in their "mother villages."

We can find many cases among the immigrants' families where a female-spouse of the family was originally from Peruvalanallur; that is, she has been married to a man in another village and, after having lived in her husband's village for some years, returned to her "mother village" with her family. The similar cases in context can be observed between Peruvalanallur and its immediate nieghboring villages, and even within Peruvalanallur.

The intra- and inter-village landownership is related to the patterns of property inheritance and marriage in the rural communities concerned. Most villagers expressed the feeling that their sons and daughters have an equal right to share their parents' property, and this practice is regarded, at least on the surface, as the basic rule for property inheritance in the studied area. However, in the actual execution of property divisions, only sons usually have an equal right to share their parents' immobile properties (residence, crop-land, fruit garden, forest, etc.), although the amount of the share might vary depending on the size of the property-holding of the individual households. In turn, for their daughters, the parents meet all expenses for their marriage, including those of the ceremony, ornaments, and cash money; and even after their marriage the parents are supposed to give cash money and gifts regularly for a long period, all of which value is assumed to be equivalent to their sons' inherited share.[3]

How can we explain the fact that many female members have registered their land properties under their own names? The important points in answer to this question are as follows:

---

3. Again, this practice could have been applicable only for 412 households (47 percent) of the total households (874) in the studied village, since the remaining 462 households (53 percent) were landless and generally poor except for a few households in each of which at least one of the working members had a decent, non-agricultural occupation.

1. Among big landlords, there is a common practice by which parents' property is given partially to their daughter(s) as part of a dowry
2. In a related manner, there is a customary rule that the mother's property brought to her husband as a dowry is supposed to be given to her daughter(s)
3. In spite of the parents' great desire to have at least a son in their family, there are quite a few families without any sons, but only daughter(s). In this case, the parents' property should obviously be given to their daughter(s)
4. There are some widows who have had a long period of married life without any issue.

Most of the female members' land in distant villages was acquired through inter-village marriage.

Size of Landholdings

The size of landholding of each household in Peruvalanallur in 1980 showed wide variation: from landless to 195.27 acres, with 1.90 acres being the average. Out of the 874 total households in the village, more than half were landless. If we take into account the landless and marginal farmers owning less than 2 acres, there were as many as 706 households (81 percent), but they together occupied only 195.14 acres (12 percent) of the total area owned by the Peruvalanallur villagers. By contrast, only 39 households (4 percent) owning 10 acres or more occupied 890.13 acres (54 percent) of the total area.

The size of the landholdings differed by caste groups. Among the 33 caste groups, the *Reddiars* (79 households) occupied 974.37 acres (59 percent) of the total area owned by the Peruvalanallur villagers, with 12.33 acres being the group's average. Among the *Reddiars*, 26 households fell under the category of holding 10 acres or more, and they together occupied 732.50 acres (42 percent) of the total area, although 4 households were landless. After the *Reddiars*, the ranking of the landholding by castes from the highest were the *Udaiyars*, (12 percent of the total area), the *Gounders* (11 percent), the *Pallans* (9 percent), and the Muslims (5 percent); and their average holdings were 1.62 acres, 2.51 acres, 0.52 acres, and 1.77 acres respectively.

A few landowning households (owning 10 acres or more) were from the "backward" caste group (11 households) and the "scheduled" caste group (2 households). In each of these castes, however, there were very high percentages of landless households (42 percent in the "backward" caste group and 65 percent in the "scheduled" caste group). Out of the 33 castes in Peruvalanallur, 14 castes had virtually no agricultural land, the members of these castes having mostly non-agricultural occupations.

The landholding of individual households and/or caste groups should not be thought of as unchangeable over a given period of years. Indeed, the above landholding patterns by size and/or caste have resulted from changes over a long period of years.

Land Transactions

Since 1864, the *Reddiar*'s percentage of landholding in Peruvalanallur has declined, with great changes occurring since 1925. In relation to this, we examined more recent changes of landholding (or transfers of *patta,* title ownership of land) in Peruvalanallur in the 12 years between 1967-68 and 1979-80.

Transfers of landownership are usually accomplished in one of two ways, (1) inheritance or (2) sales transactions. The total transfers of landholding during the 12-year period involved 365 acres in 285 cases, accounting for 39 percent of the total agricultural land available in the village. The areas which involved inheritance accounted for 192.33 acres in 184 cases, and sales transactions, 172.67 acres in 101 cases.

During the 12 years studied, transactions involving inheritance occurred in only 8 out of 33 caste groups, with great variation in the areas and cases involved among them: the *Reddiars* showed the largest involved area (132.72 acres in 115 cases), followed by the *Udaiyars* (10.97 acres in 18 cases), the *Gounders* (8.94 acres in 13 cases), and the Hindu *Pallans* (6.35 acres in 14 cases). As expected, the above figures are closely related to the number of landowning households of each caste group.

The family inheritance transaction is usually made some time after the heir's parents' (especially father's) death, with the formal registration of the transfer being made much later. However, there were some cases in which the transactions occurred even in the parents' lifetime. These were mostly a consequence of the land-ceiling acts or were the result of dowry transactions which have occurred among the limited numbers of big landowning families belonging to the *Reddiars*. The areas concerned with the land-ceiling acts and dowry transactions were respectively 57.12 acres (in 34 cases) and 15.92 acres (in 14 cases), which together accounted for about 38 percent of the total area of the inheritance transactions of the village.

Regarding property division that took place in the parents' lifetime, still another type is related to various kinds of "family troubles." Examining the individual cases under this type, they were mostly concerned with the result of unsuccessful relationships among the family members, especially between the family head and married son(s).

Regardless of the types of inheritance, these changes in landholding are the consequences of the life cycle of the individual inheritees and, by extension, of their family. A more important aspect of inheritance with regard to this study is the fact that, due to population growth within the family, the sizes of the present households' landholdings are generally much smaller compared with those of their parents' (or ancestors) as far as the inherited part of the landholdings is concerned. Thus, sales transactions are a major factor responsible for fluctuation (increase or decrease) of landholding in a given family.

During the above 12 years the sales transactions occurred in only 12 out of 33 caste groups; of these only 7 caste groups were involved in both selling and buying, 3 caste groups were involved in only selling, and 1 caste group was involved in only buying. Other villagers also were involved in the sales transactions in Peruvalanallur. There was great variation in the areas and cases in both selling and buying among the caste groups concerned. The balance of area sold and area purchased in each caste also showed a wide range of variation: among 11 caste groups involved, it was found that 5 caste groups together lost 72.15 acres of their land, the *Reddiars* alone decreasing their landholding by 70.39 acres, and that 6 caste groups together added 44.13 acres to their landholding, the Muslims, *Gounders*, *Udaiyars*, and Hindu *Pallans* increasing their land to a sizeable extent.

It should be noted that other villagers who owned land in Peruvalanallur sold 44.93 acres of land to both the Peruvalanallur villagers and people outside of the village, and that some other villagers purchased 73.25 acres of land from both Peruvalanallur villagers and outsiders. Thus, other villagers in Peruvalanallur increased their land by 28.32 acres during the period.

We realize that, during the short period of 12 years, land sales transactions occurred more frequently than one would have expected and that the areas involved were larger. It is clear that there has been a re-distribution of land among the villagers and, by extension, among the different caste groups in Peruvalanallur, the *Reddiars* as a landowning caste group having been a great "contributor." More specifically, among the individual *Reddiars* concerned in the transactions, relatively large landowning households (holding 5 acres or more in 1980) have been the main "contributors" for the land re-distribution in the village.

On the other hand, since the early 1970s, some Muslims (whose family member(s) have had jobs in some oil-producing Persian Gulf countries and have been able to amass substantial savings) have been investing their capital in land, rather than in non-agricultural business, resulting in the Muslims' higher percentage of "gained" areas among the caste groups during the period in Peruvalanallur. It should be noted that outsiders, especially those of dry villages, have had a constant interest in securing more wet land (in Peruvalanallur). Again, the Muslims in other villages have invested their "Gulf money" in buying 20.18 acres in Peruvalanallur during the 12 years studied. Indeed, the Muslims determined, to a large extent, the price of land in and around the studied village.

Thus, the Peruvalanallur villagers together in 1980 occupied 78 percent of the total privately owned land (872.70 acres) in the village,[4] the remaining 22 percent being held by outsiders, although Peruvalanallur villagers had much land in other villages.

Development of Agricultural Lands and Irrigation Systems

The present areas of "dry," "single-wet," and "double-wet" lands in Peruvalanallur are a consequence of the expansion of irrigation. In relation to the development of agricultural land, the *Settlement Registers* (published in 1864, 1898, and 1927) and their corresponding cadastral maps provide valuable materials. These materials and the current data collected by the author together show how agricultural land has changed in relation to the development of irrigation systems over the period of the years examined.

In Peruvalanallur in 1864, of the total agricultural land (1,048.56 acres), dry and "single-wet" lands occupied 480.53 acres and 568.03 acres respectively. Although the Peruvalai channel appeared first in the index map of the *First Re-settlement Register* (1898), it is believed to be one of the early irrigation channels developed by the Chola Kings in the Cauvery delta area, and the channel itself was constructed around the second century, A.D.[5]

---

4. The resident *Reddiars* in Peruvalanallur occupied, in 1980, 510.44 acres, or 58 percent of the total privately owned land in the village.

5. Shanmugam P. Subbiah, "Rural Base in a South Indian Village: A Study into Its Structural and Spatial Patterns in Mahizambadi Village of Tamil Nadu," *Studies in Socio-cultural Change in Rural Villages in Tiruchirapalli District, Tamil Nadu, India,* no. 4 (Tokyo: ISLCAA, August 1981), p. 21.

Thus, by 1864 Peruvalanallur already had its own irrigation systems based on waters coming mostly from the Cauvery/Coleroon River, but partly from local water sources, stored in the two large tanks through the Peruvalai channel. However, the Cauvery River and its distributaries showed great seasonal fluctuation in the volume of water carried, there not being any reservoirs in the upper streams at that time. As there were only a few field canals observed in the "single-wet" area, the distribution of water must have mainly been carried out on a field-to-field basis.

During the period 1864-1898, 150.44 acres were converted to "single-wet" land, this being accompanied by construction of the field canals at the village level. The distribution patterns of the field canals in 1898 was fundamentally the same as the current one. Moreover, by 1898 individual areas had been assigned for irrigation from one of the two tanks, the *mela eri* or *kila eri,* or from the Peruvalai channel. Interestingly, this arrangement still exists.

During the period 1898-1927, 120.48 acres of dry land were converted to wet land, and 72.82 acres of "double-wet" land were created mainly by the conversion from "single-wet," but partly too from dry lands. New field canals were observed in the newly-converted "single-wet" areas. The southern fields had been almost completely transformed to wet land by 1927.

During the 52 years between 1927 and 1978-79, 87.48 acres of dry land were further converted to wet land, mostly in the narrow strip immediately to the north of the Peruvalai channel. More importantly, 508.4 acres of "double-wet" land were newly created by converting mostly "single-wet," but also some dry lands. It should be stressed that the fundamental factor contributing to this remarkable expansion of the "double-wet" land must have been the Mettur Dam of the Salem District (located some 178 km northwest of the studied area), the construction of which was completed in 1934. The large-scale reservoir behind the Mettur Dam has been well coordinated with the Upper Anicut (located some 20 km west-northwest of Tiruchy), where waters for the Peruvalai channel, and other channels as well, are controlled by the Public Works Department of the Tamil Nadu Government. Most of the waters in the Peruvalai channel are stored in the two large tanks available (which waters are eventually distributed through its several outlets to the different sections of the southern fields), but some enter directly to the main field canals which run in a north-south direction in the studied village.

In relation to the expansion of "double-wet" land, it should be noted that some innovative farmers recently have installed shallow tubewells (with diesel engine pump-sets) in the southern wet-lands and deep tubewells (with electric pump-sets) in the northern "dry" land in the studied village.

Land-Use Patterns

In the studied village, which belongs to a part of the "upper wet" zone in our physiographic model, the agricultural areas are mostly wet land, but they also contain some higher "dry" land, including uncultivated land.

Major crops in the wet land are different types of paddy *(kuruvai, thaladi,* and/or *samba)* and sugar cane, although various types of gram and other minor field crops are extensively cultivated in the beginning of the dry season in the same fields used for the major crops in the wet season. Since sugar cane (and bananas) takes almost one year to harvest, their area has been regarded officially in the classification of land as a "double-cropping area," like that of the *kuruvai* and *thaladi* paddy fields. On the other hand, the *samba* area has been officially treated as a "single-cropping area" because there is no other major crop cultivated during that time of the year. It should be pointed out that the traditional categories of "single-cropping" and "double-cropping" include gram and field crops to a lesser or greater extent.

We found in 1979-80 the following types of crop associations in wet land in Peruvalanallur:

1. Single-cropping in wet land . . . . 171.35 acres
   a) *samba* paddy
   b) *samba* paddy - gram
2. Double-cropping in wet land . . . . 650.70 acres
   a) *kuruvai* paddy - *thaladi* paddy } (325.87 acres)
   b) *kuruvai* paddy - *thaladi* paddy - gram }
   c) sugar cane } (342.83 acres)
   d) sugar cane - gram }

Land use in "dry" land on higher elevation is more diversified in spite of its relatively small area. This is not only because of the physical differences within the "dry" land itself, but also because the farmers' responses to the land are varied. This character is an attribute of the the intermediary position between wet land and *real* dry lands. Thus, we found the following types of crop associations in the "dry" land in Peruvalanallur:

1. Single-cropping in dry land (only rain fed): ground nuts, red gram, maize, black gram, field beans and ragi *(Eleusine coracana)* . . . . 31.31 acres
2. Fruit trees in "dry" land (with irrigation) . . 13.33 acres
3. Single-cropping in "dry land (with irrigation)
   a) *samba* paddy . . . . 15.20 acres
4. Double-cropping in "dry" land (with irrigation) 12.64 acres
   a) *kuruvai* paddy - *thaladi* paddy . . . . (6.84 acres)
   b) sugar cane . . . . (5.62 acres)

Cultivated areas of individual crops over the different seasons of the year strongly reflect the year-round agricultural activities of Peruvalanallur villagers. Periods of planting and harvesting the crops are the two busiest seasons, the latter being much busier than the former. However, in the wet zone, where double-cropping is dominant, harvesting of the *kuruvai* paddy and the planting of *thaladi* paddy and/or *samba* paddy must be completed within the limited period from the middle of September through the end of October. This is due to the northeast monsoon rains which, while having a negative influence on the *kuruvai* harvest, are a positive factor in the entire process of the *thaladi* and/or *samba* paddy cultivation and standing sugar cane. Indeed, in this period in the wet zone of the Lalgudi Taluk as a whole, the harvesting of *kuruvai* and preparation of the fields and transplantation of *thaladi* and/or *samba* can be observed simultaneously. Agricultural work in this period is more intense in the "lower wet" zone than in the "upper wet" zone, since in the former double-cropping of *kuruvai* and *thaladi* is almost exclusively practiced. Another busy season is the period from early January through the end of March, during which *thaladi* and/or *samba* and sugar cane are harvested.

Recent Change in Agricultural Practices and Technologies

The agricultural practices and technologies employed in the studied village reveal some important characteristics. In Peruvalanallur, almost all farmers currently employ the high yielding varieties (HYVs) for each type of paddy, sugar cane, gram, and other field crops. As is mainly the case in the wet zone in Lalgudi Taluk, the introduction in Peruvalanallur of the HYVs for each type of paddy in particular began in the middle of the 1960s (corresponding to the beginning of the so-called "Green Revolution"), almost completely replacing traditional varieties by around 1975. Delayed by only a few years, sugar cane, gram, and other field crops followed nearly the same process as was the case for paddy. It should be pointed out that the newer HYVs for the individual crops have repeatedly been replacing the older HYVs. Adoption of other modern inputs, such as chemical fertilizers, pesticides, etc., are commonly practiced among the villagers, although some poor farmers' investments are slightly lower. For these agricultural inputs for each of the major crops, the government-linked Co-operative Bank provides the credit for actual cultivators, this having to be certified by the village *karnam* (caretaker of the land record).

Farming technology in general has been associated with traditional implements and tools, except for the pesticide sprayers and tractors. In 1968, the first tractor (35 HP) was introduced in the village by a *Reddiar* who, in plowing others' fields or earning cartage for sugar cane, paddy, and other agricultural goods, used this first tractor mainly for "business," rather than for his own agricultural purposes. In 1980, there were three tractors (each 45 HP) owned by the *Reddiars* in the village. Since the first adoption of the tractor in 1968, the area in the village cultivated by tractors has gradually increased. In 1980, the area cultivated by tractors accounted for about 70 percent of the total cultivated area,[6] the rest by either bullocks or buffaloes.[7] As a consequence, the numbers of work cattle and water buffaloes used for plowing have been greatly reduced in the village. Efficient plowing by tractors has been especially important in the period from mid-September through the end of October, during which, after the *kuruvai* harvest, subsequent preparation of the fields for *thaladi* and *samba* paddy has to be completed within a limited number of days. The tractors have also been used for threshing of the paddy, although beating at the stone plate and/or cattle-treading methods are more popular in and around the studied village.

Human and animal sources of energy are still predominantly used in operational aspects of farming, but this does not greatly limit agricultural production. Rather, it helps wage laborers maintain their employment. However, when more efficient work is necessary, farmers employed modern implements, such as in their use of the tractor.

---

6. Total cropped area represents the aggregate area of all crops raised on the same land during the different seasons of the year.

7. This figure is not necessarily indicative of a limit in the tractors' capacity, but rather tractors did not have access to some fields which were either blocked by physical obstacles or surrounded by already transplanted paddy fields.

Selection of Crops and Recent Change in Land Use

Selection of the specific seasonal crops for particular fields depends not only upon the physical conditions needed by the individual crops, but also upon the differing production values among the crops. Under the current irrigation systems, two types of paddy (*kuruvai* in the first season and *thaladi* in the second season), as one set, and sugar cane are, in most "double-wet" areas, usually interchangeable, although there are some slight differences in the physical conditions required by each. It should be recalled that, after two consecutives years' harvests, the sugar cane is replaced by the other crop (paddy) because of its gradual diminishing yields from the first year to the second and to the third. After the cane's replacement, the same land can be used in two ways in the following seasons: (1) *kuruvai* paddy cultivation in the subsequent first season of the year and (2) skipping the planting of any major crop in the first season, following with *samba* paddy cultivation in the second season.

Whether or not sugar cane (or paddy) should be cultivated in particular fields during the year largely depends upon the results of the farmers' "benefit-cost analysis," based on experience. In recent years, the market price of processed sugar has been increasing, and accordingly sugar cane has become the most profitable crop. Thus, there has been an increasing interest in sugar cane cultivation. In fact, the cultivated area for sugar cane increased by 41 percent in the year between 1978-79 and 1979-80. Many farmers converted from *samba* paddy in 1978-79 to sugar cane cultivation in 1979-80: this means that farmers converted land from a "single-cropping" to a "double-cropping" area within that year. From crop associations stated above, the conversion from *samba* area to sugar cane area in the next year was possible if the *samba* cultivation in 1978-79 was the consequence of crop associations following sugar cane harvest in the previous year (the land already being double-cropping area).

However, this only partly explains the fact that of the area (95.93 acres) converted to sugar cane in 1979-80, 21 percent came from the "double-cropping" area and 79 percent from the "single-cropping" area of 1978-79. In the full explanation of the above fact, two other elements are thought to be important: (1) some farmers could secure subsidiary irrigation water for sugar cane cultivation, this being provided by the nearby shallow tubewell owners on a contract basis; and (2) some farmers, whose fields were adjacent to others' sugar cane fields, too desperately cultivated sugar cane, hoping that their fields could secure some irrigation water through their neighbors' fields. It should be pointed out that there was much more area falling under this second type than under the first. As expected, however, there were great differences in average yields between the two types: the yields of fields with "borrowed irrigation" were only half those found in fields with contracted irrigation. The farmers' crop selection in both types of irrigation depends on other neighboring farmers whose fields are of better quality and/or equipped with the irrigation facilities. Though land cultivated under the above two methods can be regarded as temporary double-cropping land (being cultivated in this way for only one year and having generally low yields), it certainly has the potential to become a more stable cropping area. It should be noted that even the core part of the current double-cropping area had, for a long period of time, been developed by similar kinds of "trial and error" processes before taking the shape in which it is found today. Thus, we classified the wet land into three category areas based on its uses for the "single-cropping" and "double-cropping" in the two consecutive years of 1978-79 and 1979-80 as follows:

1. Area for stable single-cropping . . . . . . 118.41 acres
   a) single-cropping area in both years
2. Area with double-cropping "potential" . . . 135.74 acres
   a) double-cropping in first year, then single-cropping in second year . . . . . (52.94 acres)
   b) single-cropping in first year, then double-cropping in second year . . . . . (82.80 acres)
3. Relatively stable double-cropping area . . 569.42 acres
   a) double-cropping in both years

What should be the farmers' great concern in the future is the promotion and maintenance of the relatively stable double-cropping land and its "potential" to become more stable, productive land, converting into a double-cropping area the remaining one-fourth of the total wet (lower) land, currently used as "single-cropping" area for *samba* paddy. As availability of irrigation water is the key factor for these achievements, farmers must take the initiative to secure more local water, emphasizing the need for the installation of more irrigation facilities, such as shallow and deep tubewells in the southern fields in the village.

Agricultural Labor Force and Its Organization

Peruvalanallur village in 1980 had 1,200 agricultural wage laborers (540 males and 660 females), of whom landless wage laborers were 635 persons (297 males and 338 females). In spite of the availability of such a large number of wage laborers in Peruvalanallur, during the busy seasons the area receives many laborers from outside the village, especially from dry villages. Although the seasonal migration from the dry villages is not a new phenomenon, it has been greater in recent years, especially since the second half of the 1960s. Many seasonal wage laborers come to the village without previous arrangements and ask work of "would-be" employers. Some big landowning *Reddiars* procure groups of laborers (15-20 persons per group) on a contractual basis each season through their relatives in the dry villages. Although the wage of the seasonal laborers from outside the village generally is lower than that of Peruvalanallur laborers, it is at least 1.5 times that found in their own dry villages.

In connection with the seasonal migration of the wage laborers, the recent development of double-cropping in the wet zone has had general impact not only on its own labor structure, but also on that of the less-developed dry zone, resulting in a seasonal shortage of labor there and, by extension, changes in farm management procedures, including the introduction of *varam* (share-cropping).

Regardless of the size of the landholdings, managing members within the families usually are responsible for agricultural operations. The major physical work for agricultural activities on the individual farms, however, depends largely on the hired laborers, this being the case even in the marginal units, which have such work members in their own families. Most of the manual work for agricultural activities is conducted intensively and collectively in a limited period on the individual farms by employed laborers.

There are two common wage systems for the agricultural day laborers: (1) the ordinary fixed wage system and (2) the contract labor system. Some types of work are assigned only to male laborers, others only to female laborers. The agricultural laborers in and around the study area in 1979-80 worked 5 hours a day, including a morning time (9:00-12:00) and an evening time (4:00-6:00). Male laborers were paid Rs. 5 (Rs. 3 for the morning time and Rs. 2 for the evening) per day, whereas females were paid only Rs. 2.25 (Rs. 1.50 for the morning and Rs. 0.75 for the evening). This wage system applies to the various kinds of work done, except for mainly paddy and sugar cane harvests. Only for paddy harvesting is an equal wage in kind (i.e., paddy) given to both male and female laborers.

The contract labor system is applied to every stage of paddy and sugar cane harvesting, as well as to some other specific jobs such as plowing, transplanting of paddy, spraying of pesticides, application of fertilizers, etc. Both male and female laborers in a party under the contract system usually enjoy a higher wage. This is a result of the collective bargaining power of the laborers in the system, itself a consequence of the time schedules which the cultivators must meet, especially during the two busy seasons. It should be noted, however, that the individual laborers in this system not only achieve more work per hour, but also extend their working time by 1-3 hours per day over those working in the ordinary fixed wage system.

Whichever wage system is employed, the cultivators have to meet the substantial costs involved in supporting the labor force, this being indicative of the labor-intensive character of the agricultural systems.

It should be pointed out that, except for the "forward" castes, all the sizeable caste groups have considerable numbers of agricultural wage laborers and the "working party" in the contract labor system consisted of members from varied castes. This indicates that, at least for the purpose of economic gain, there was a common base of association and/or integration among agricultural wage laborers whose caste backgrounds are different. A drastic example of the power of this common base was seen in the "wage laborers' strike," occurring early in October of 1980 in Peruvalanallur and initiated by the *Harijan*s demanding that the employers (*Reddiars*) increase their daily wage. Although there have in the past been frequent instances of conflict between the *Reddiars* and the *Harijan*s, this was the first occasion in village history that the laborers held a village-level strike, with the "backward" caste groups acting jointly with the *Harijan*s against the *Reddiars*. Though this began for the economic gain of the laborers, it obviously indicates some aspects of social change and/or a re-structuring of the social relations among the varied caste groups in the rural community.

Land Tenure

Under the current agricultural practice and technology and the cultural configurations of the rural communities, the size and amount of means of production, such as land, labor, agricultural implements, and livestock (especially bullocks as draft power), were unevenly distributed among households in the given village. Accordingly, each household would try to utilize its available resources as effectively as possible in accordance with these differences. Representative means of such effective utilization of available human and material resources are seen in the participation in various kind of tenancy.[8]

There are three types of land tenure currently available in Tamil Nadu; namely, (1) *varam* (share-cropping tenure), (2) *kuttagai* (fixed rent tenure), and (3) *otti* (usufructuary mortgage tenure). However, all types do not occur equally in all the villages, *taluks,* or districts. Commonly only one type or two are dominant in a given village or area. It should be stressed that, since the 1950s in Tamil Nadu State, the legislature has passed several tenancy acts and rules, the amendments to which are aimed at protecting the "cultivating tenants," including any member (heir) of their famiy, and that all types of land tenure have been influenced by these acts and rules to a lesser or greater extent, this being the case even though awareness of such acts varies among villagers involved.

Although *varam* is not practiced at all in Peruvalanallur, it was found in some dry villages in Lalgudi Taluk where some modern irrigation systems have been developed in recent years. In the current *varam* system, the amount of rent is determined by the total gross produce of the involved land and share-ratios betwen the landlord and his tenant. The share-ratios of return between the landlord and his tenant vary greatly, depending not only upon the different provisions for irrigation water for the involved lands, but also upon the cost-bearing conditions between the two parties for kinds of agricultural work, fertilizers and pesticides, seeds, and operational work on and maintenance of the irrigation facilities.

In some dry villages of Lalgudi Taluk, modern irrigation methods have been emphasized by innovative landowners: (1) many deep tubewells with electric motors have been installed in new places, (2) some of the traditional wells have been converted into motor-pumped tubewells, and (3) quite a few low-lift pumps have been

---

8. Yoshimi Komoguchi, "Rural Community and Agricultural in Bangladesh," *Science Report of Geography,* (Tokyo: Komozawa University, March, 1982), pp. 64-65.

introduced, replacing the traditional *aetram-eravais* and *kavalais*. In a dry environment the farming of *nanjais* (irrigated wet lands) with the new irrigation systems, accompanied by modern agricultural inputs, provides much higher yields than the *punjais* (nonirrigated dry lands) or *nanjais* using traditional irrigation systems. This requires, however, more capital and a larger labor force. The landowners' varied labor situations are primarily responsible for the decision of whether or not to enter the *varam* tenure, as well as the degree of their involvement in the *varam* farming when such a decision is made. In recent years the labor force available in dry villages has been subject to great seasonal variation. This is closely related to the seasonal migration of its wage laborers to the wet villages, the demand for a labor force in these villages being much higher than that in the dry villages. Therefore, the landowners whose farming largely depends upon an outside labor force have to seek some way to obtain a stable year-around labor force. One way to solve the problem for such landowners is to choose the *varam* (share-cropping) tenancy by providing some reasonable conditions for their tenants.

The landowners are afraid that their tenants will gain a "permanent right of cultivation" to the leased lands under the 1969 Tenancy Act. Under the *varam* tenure, the landowners can defend themselves from the tenants' claims, should they occur, by insisting that the landowners have been managing the farms themselves.

In the studied village, Peruvalanallur, only *kuttagai* (fixed rent tenure) and *otti* (usufructuary mortgage tenure) are observed and commonly practiced. The fact that the Peruvalanallur villagers own more acreage in other villages (59 percent) than in their own village (41 percent) reflects its intra- and inter-village land tenure transactions. In 1980, there were 310 individual households (35 percent of the total village) involved in the two types of land tenure (165 households for *kuttagai*; 210 households of *otti*, of which 65 households practiced both tenancies). Of the total land owned by Peruvalanallur villagers, 300.27 acres, or 18 percent, were leased under either *kuttagai* (205.24 acres) or *otti* (95.03 acres).

The Hindu temples and other religious organizations of Peruvalanallur and outside village temples owned wet lands in the studied village, leasing them out to the Peruvalanallur tenants under the *kuttagai* tenancy. Outside village and town dwellers also have *kuttagai* relations with the Peruvalanallur tenants, the lands involved being located in and around the studied village. About one-fifth of the arable land of the village is not cultivated by the landowners, but by the tenants under either the *kuttagai* or *otti* systems.

Kuttagai seems to be the most common type of tenure in the Tiruchy District. Under the *kuttagai* system the tenant is supposed to pay a fixed rent in cash or kind to his landlord, the amount being settled upon before cultivation, and the tenant has to bear all the cultivation expenses. In the wet villages where different types of paddy, sugar cane, and banana are the major crops, the rent is paid in kind for paddy cultivation and in cash for sugar cane and banana cultivation. In the dry villages, cultivable lands are mostly *punjais* (nonirrigated lands) which are used for several types of pulses and millets, groundnuts, chillies, vegetables, etc. The rent for such lands (*punjais*) is usually paid in cash regardless of the kinds of crops cultivated.

The fixed rent per unit of area varies greatly not only within a village but also in the different villages or regions in accordance with the various qualities of the *kuttagai* lands, which provide different land productivities. In wet villages like Peruvalanallur, rent is determined largely by whether the land is characterized as "single-cropping" or "double-cropping" land. It is arguable whether or not the current amount of fixed rent is high (or low). It has, however, certainly become lower than in the past, the productivity per unit area having increased greatly in the last 7-10 years. Another important aspect for the tenant is that he can secure extra work (and wage income) by participating in the *kuttagai* tenure. For instance, during the paddy and sugar cane harvest, a time during which the contract wage system is especially popular, the tenant can become a member of the "work party" for his operating (*kuttagai*) land and thus get an equal share as a working member.[9]

It should be noted that, in many cases in the *kuttagai* transaction, there are differences between the stipulated amount and the actual amount of rent payment, for most of the tenants "bargain" with the landowners after the harvest.

It is generally true that the available *kuttagai* lands in a given village, regardless of their ownership, were cultivated mostly by the tenants of the same village or by the neighboring villagers. This indicates that distance is an important factor in the practical operation of cultivating *kuttagai* lands. The fact that the Peruvalanallur tenants have actually leased wet lands almost exclusively can be explained by this factor too, since the village

9. Thus, for example, the "gross produce" of paddy excludes some grain, this being paid in kind to the wage laborers for their work in its harvest.

itself lies in an extensive region with a wet environment. However, the distance factor depends upon other agricultural conditions in a given village or area. The farmers in dry villages (where agricultural activities are relatively stagnant) usually commute to fields more distant from their homesteads for farming than do those in the wet villages.

The crop yields per unit of area in the wet land in a wet environment like Peruvalanallur are generally more than three times those in the dry land in a dry environment, although there is a great variation within each of the wet and dry lands in a given milieu. This is another reason why most of the Peruvalanallur tenants desire to cultivate wet rather than dry lands whenever possible.

We examined the households involved in the *kuttagai* tenancy in relation to the caste and size of landholding. Among the 33 caste groups in Peruvalanallur, only 16 groups were involved in the *kuttagai* tenancy (landowner-side only: 4 groups; tenant-side only: 7 groups; and both sides: 5 groups). The *Reddiars* (16 households) leased 174.33 acres or 85 percent of the total *kuttagai* lands (205.24 acres) owned by the Peruvalanallur villagers. The other caste groups who leased sizeable areas were the Muslims (9.36 acres), the *Udaiyars* (8.85 acres), and the *Gounders* (8.15 acres). Each of the remaining 5 caste groups respectively had one household involved in the *kuttagai* tenancy, and its leased land was a very small area (1.50 acres at most). On the other hand, among the 12 caste groups of the *kuttagai* tenants, 5 groups (*Reddiars*, *Udaiyars*, *Gounders*, *Muthurajas*, and Hindu *Parayans*) were the most important ones as far as the involved households and their leased areas are concerned.

As far as the number of the involved landowners' households in the *kuttagai* tenancy is concerned, there was no relation to the size of landholding: there were 17 households belonging to the marginal landholding group (owning less than 2 acres). Most of these households leased all of their lands. However, since their respective sizes of landholding were small by definition, the total leased area and area per household were accordingly small. In general the larger landowning households leased more area per household. However, it should be noted that the largest landowning household *(Reddiar)* in the studied village alone leased as much as 136.41 acres of their land. More specifically, one household alone leased 12.24 acres (wet land only) in 18 different contracts with the Peruvalanallur villagers (who belonged to 7 different caste groups), and 124.17 acres in 67 different contracts with the other villagers. This explains the extremely high ratio of the leased area by the *Reddiars* and by the highest landowning class as well.

As would be expected in the *kuttagai* tenancy, the landless and marginal landowning households are more involved in the tenancy. Out of the 129 tenants' households in Peruvalanallur, 90 households or 70 percent consisted of the landless (44 households) and marginal ladnowning groups (46 households), and they together leased 77.00 acres or 53 percent of the total *kuttagai* area (145.28 acres) leased by the village tenants. It is generally true that the smaller landowning households leased a smaller area per household. This is certainly related to the tenants' capability for farm management, as the tenants in the *kuttagai* tenancy have to meet all expenses for cultivation.

With regard to the caste relations in *kuttagai,* a landlord usually chooses his tenant(s) either from his own caste or from socio-economically lower-ranking castes. Conversely, a tenant seeks his landlord(s) either from his own caste or from socio-economically higher-ranking castes. This is certainly related to the barrier of the caste psychology. The one exception is the Muslims, who do not have such cultural barriers.

Another important aspect in the landlord-tenant relationship under the *kuttagai* tenancy is that, in a given contract, the size of the landholding of the landlord is not necessarily larger than that of his tenant. The implication of this is that, under such landlord-tenant relationships, the landlord was not necessarily more influential, socio-economically speaking, than his tenant. With regard to landlord and tenant relations, our data reveal that the majority of the current *kuttagai* contracts have continued over long periods, that is, 80 percent of the total contracts have continued for more than 10 years. The fact that the recent *kuttagai* contracts, which have started within the last 10 years, make up a small percentage of all *kuttagai* contracts (based on data as of March 1980) is certainly related to the influence of the 1969 Act. It should be noted that many of the recent contracts were among close kinship groups. How should we take into consideration the fact that the great majority of the counterparts in the current *kuttagai* contracts have not changed for a long period of years? One way to look at this fact is that the landlords might not have been as strict with their tenants as is generally understood, although there have been exceptions.

Under the current *otti* (usufructuary mortgage tenure) system, a tenant gets the right to cultivate the land involved by depositing a certain amount of cash in advance with his landlord. The period of the *otti* contract is usually for three years, but it can be renewed if both parties (landlord and tenant) agree on its new terms. The full right of the land involved is returned to the landowner on the repayment of the deposit without any interest to the tenant. Thus, we can regard the tenant's (creditor's) yearly enjoyment of cultivating the *otti* land as an "annual interest" on the cash deposit to the landowner (debtor).

Out of the 874 households in Peruvalanallur in 1979-80, 210 households or 24 percent were involved in *otti* tenancy; of these, 94 households were landowners (debtors) and 124 were tenants (creditors), although 8 households were involved in both. Like the case of the *kuttagai* tenancy, both landowners and tenants of Peruvalanallur had *otti* transactions not only with co-villagers, but also with outside villagers, and vice versa for outsiders.

The 94 households in Peruvalanallur together received Rs. 636,270 of the cash "credits" in exchange for the leasing of 95.03 acres under the *otti,* these being located not only in Peruvalanallur (69.74 acres), but in its neighboring villages as well (25.29 acres). Of these available *otti* lands leased by the Peruvalanallur villagers, the tenants of the same village cultivated 80.19 acres, and the outside village tenants cultivated the remaining 14.84 acres.

On the other hand, the tenants of Peruvalanallur together leased 91.90 acres of *otti* lands not only from their own village (80.19 acres), but also from the other village and town dwellers (11.71 acres), paying Rs. 656,270 for the right to cultivate the areas involved.

At this stage, some important aspects should be pointed out. First, out of the total area involved in the *otti* tenure in the studied area in 1979-80, the ratio for wet land was extremely high compared with that for dry: comparing the *otti* and *kuttagai,* it was higher in the former (86 percent) than in the latter (75 percent). Second, unlike the case of the *kuttagai* tenancy, out of the total *otti* areas leased from the Peruvalanallur landowners, the other village tenants had a very small share compared with that of the Peruvalanallur tenants. Third, the spatial transactions of the *otti* between Peruvalanallur and the other villagers were limited in comparison with those of the *kuttagai.* This is certainly related to the fact that the wet lands generally provide higher productivity under relatively stable conditions for cultivation. Careful farm management by frequent visits has also become more important for the successful farming of these wet lands because of its capital-labor intensive character.

The amount of "credit" for the *otti* land was settled primarily on the quality of the land involved. For the wet land the amount of "credit" varied widely, depending upon whether the land was a "single" or "double" cropping area, and for the dry land, whether it was equipped with irrigation facilities or not.

It should be noted that all the types of crop associations in Peruvalanallur also were observed in the *otti* lands, because the *otti* lands in the studied village were distributed almost evenly regardless of the different physiograhic conditions of cultivation. When the tenant's profits were examined by taking into account all the available crop associations in the studied village, the obtained figures revealed high rates of imputed interest with a range of 30-55 percent per annum. Although these rates cannot be claimed to be perfectly accurate, they correspond well to the villagers' general understanding that the *otti* tenants can safely get back one-third of the amount of "credit" given to their landowners in a year. Realizing that a favorable rate of annual interest for rural people depositing in the authorized banks was 10-15 percent, it is true in a sense that the tenants' involvement in the *otti* tenure can be regarded as a highly productive investment.

The results of interviews indicated that the *otti* tenancy has been practiced for a long time in and around Peruvalanallur. However, the number of transactions and the area involved in the *otti* system have predominantly occurred only after the late 1960s; since then, the *otti* transactions have become increasingly popular among the villagers. The beginning of the popularity of the *otti* transactions corresponds to that of the so-called "Green Revolution" in and around the village. In other words, the *otti* transactions have been emphasized in accordance with the processes recently implemented in the development of its agriculture.

The rapid rise of *otti* transactions may be related also to the stagnation in the *kuttagai* market, which has been largely closed to new transactions in recent years, mainly because of the strong influence of the 1969 Tenancy Act. However, the major reasons for the prominence of *otti* transactions in Peruvalanallur can be found in the recent rise in the productivity of agriculture itself.

The Changing Village

The development of agriculture in and around the studied village has been largely dependent upon that of irrigation systems over a long period of years. There were noticeable characteristics in the development of irrigated lands during the period 1864-1980: (1) an extensive conversion from dry to "single-wet" lands during the relatively long period 1864-1925, in which the current network of field canals had been fundamentally shaped by 1898, and (2) a remarkable conversion from dry and/or "single-wet" to "double-wet" lands during the period 1927-1980, wherein intensive conversion occurred in the 1930s associated with construction of the Mettur Dam, and again in the second half of the 1960s and onward, associated with the private installation of shallow and/or deep tubewells.

On the assumption that the above process of development of agricultural lands and, by extension, growth of agricultural production might have occurred in most parts of the alluvial plains of the Cauvery River basin (although there would have been some regional variation within it, as observed in its current situation), there must have been two periods of so-called "Green Revolution" since 1927. There also is a possibility that another "Green Revolution" occurred in some years during 1864-1898, when the marked development of field canals at the village level took place during a relatively short period of years.[10]

When one focuses upon the most recent "Green Revolution" observed in the second half of the 1960s on onward, agricultural activities have become more vigorous than ever before. It has much emphasized modern agricultural inputs and the adoption of improved practices and technologies, along with credits through government-linked cooperatives at the village level.[11] Thus,

---

10. In any given rice-producing region or country, the development of various levels of man-made canals is one of the best expressions of the developmental stages of agriculture.

11. There is wide spread speculation in both the academic and public spheres that the institutional credit for agriculture has mostly been assigned to the big landowners, neglecting the small farmers. It is a fact, however, that the amount of agricultural credit for individual households in general has been allocated through the Village Cooperatives based on the per unit area of their operational landholdings.

Since most credit has been sanctioned through the Village Cooperatives, farmers' easy access to institutional credit is dependent upon the availability of well organized cooperatives. It should be stressed, however, that there is a great deal of difference in their organization and activities. At established village cooperatives like the one in the studied village, farmers can get credit without much difficulty, regardless of the size of their landholdings. This is because the village cooperatives, being autonomous bodies, have benefited in accordance with the amount of transactions in agricultural credit, and thus they tend to provide the farmers (share-holders of the cooperatives) with a maximum amount of credit within the limit of official allocations. Since most of the interest rates on loans are as high as Rs.12-15 per annum, only the farmers who had carefully utilized credit for better farming wuld achieve the expected economic gains. Unfortunately, some farmers have not used the credit for what

agricultural productivity has been increasing; the apparent "greening" has been through the intensification of cropping. Farming has become more capital- and labor-intensive in character, and thus farm management has become more important for successful production.

Adoption of improved technologies is one of the key factors for modernizing agriculture. The agricultural practices and technologies employed in and around the study area consist of "traditional" and modern ones in which the former is associated with human and animal sources of energy in its operation and the latter with inanimate energy. It should be noted that the currently employed agricultural practices and technologies have resulted mostly from the indigenous development of integrated efforts by intelligent and/or innovative villagers (who are not necessarily farmers),[12] extension specialists, agronomists, irrigation engineers, other related scientists and technologists, and government specialists under the development agencies. It is a fact that the traditional practices and technologies are still dominant, but there have been well coordinated with the modern ones. In general, adoption of modern technology in rural India has been exercised only when agricultural production would be considerably reduced without it or, conversely, be remarkably increased with it.

The legal control over land tenureship is another responsible factor for modernizing agriculture. It should be recalled that since the later 1950s the legislature of Tamil Nadu has passed several tenancy acts and rules aiming primarily at protecting the "cultivating tenants," of which the 1969 Act (Tamil Nadu Agricultural Lands Records of Tenancy Act) has been the most influential. The recent rapid increase of agricultural production has proceeded under the strong influence of the various tenancy acts and rules.

Enforcement of legal controls has led to the following major consequences: (1) it has shaken up the existing landlord-tenant relationships in *kuttagai* (fixed rent tenure) by which some tenants have purchased the lands involved at a reduced rate and other tenants have returned the lands to their landlords with receipt of some cash money; (2) the new *kuttagai* markets have almost been closed; and (3) the landowning households (mostly *Reddiars*) who are

---

they ought to have, and thus they have accumulated outstanding principal and interest over the years. In order to clear the loans from the cooperatives, quite a few landowners have pledged their lands under *otti* systems.

12. This includes job-seeking college and university graduates who usually stay in the villages and help in the managerial part of farming.

short of working members in their families have been reluctant to expand their landholdings but have preferred to invest their capital in the non-agricultural sector including agro-based business. As far as the growth of agricultural productions is concerned, these changes themselves have generally worked out positively, although there arguably remain problems of social justice in the process of tenancy resolution.

The changing agricultural systems have interacted with the socio-economies of the individual households and village communities as a whole. Whereas agricultural productivity has been increasing in the recent years, some households could not adapt themselves to the new conditions of farming and to the overall socio-economic changes in the village communities, and thus they have had to face the constraints of family maintenances and farm operation itself (if they ever have had lands). As part of the immediate solutions for these difficulties, considerable numbers of households have had either to sell a portion or all their individual holdings of land, or to pledge their lands under *otti* (usufructuary mortgage tenure), or be involved in both land sales and *otti* transactions.

Of course, there have been various reasons for individual landowner's involvement in land sales and *otti* transactions. Similar reasons (and motivations), however, can commonly be found both in land sales and *otti* transactions as follows: (1) family constraints, (2) consequences of the various influential tenant acts, (3) less productive lands (dry and single-cropping lands) or difficulty of farm manangement of the involved lands located in distant villages, an (4) creation of capital for non-agricultural business.

Regarding the *otti* transactions, the capability of the landowners' clearance of indebtedness (and re-enjoyment of the full right of cultivation of the land involved) is, of course, entirely dependent upon their individual economic achievement thereafter. Since the current agricultural productivity in general is high and stable, the *otti* landowners' economic achievement can be measured to a degree on the percentage of leased area of the individual landowner's total holdings. If the leased area was relatively small, the landowner usually could get back the *otti* land, but if the leased area was relatively large (or the involved loan was large),[13] the landowners would, after all, have had to sell a

---

13. Such cases in general were the outcome of repeated *otti* transactions of the same lands in every three years by raising some amount of "credit" per unit of area in each new contract and/or expanding to other lands under *otti*.

portion or all of the lands involved.[14] These cases exist regardless of the size of holdings and of caste groups, but are prominent among the landowning *Reddiars*.

The recent rise in the productivity of agriculture has induced the prominence of *otti* transactions in which a considerable number of landless and marginal households have been involved as tenants (creditors). As far as the growth of agricultural production is concerned, the popularity of *otti* transactions is assumed to have worked out positively, because the *otti* tenants are apparently better equipped for meeting the new agricultural conditions.

The impact of the "Green Revolution" on agricultural labor and its organization also has been an important aspect. The "Green Revolution" in rice production has induced more work per unit area and has thus provided more jobs for agricultural laborers, since, unlike the wheat-producing areas, it has proceeded without much emphasis on the adoption of labor-saving improved implements and modern machines.[15]

Corresponding to the growth of agricultural productivity, laborers' wages have increased in almost every year. It should be noted, however, that the recent trend of laborers' intensification of day work has been reflected in wage increases to some degree. Related to the recent increase of agricultural work per unit area, labor employment under the contract system has bee increasingly popular compared with that under the ordinary fixed-wage system.

It is not a new phenomenon that some different sub-caste members of the "Backward" or "Scheduled" classes work together under the contract system. In recent years, however, this has become more prominent as more varied sub-caste members have been organized into working parties regardless of the differences between the "Backward" and "Scheduled" castes. The timely organization of proper numbers of day laborers has been increasingly important for both landowners and laborers. This has induced more frequent and direct association not only among the wage laborers, but also between landowners and laborers with different caste backgrounds. The intensification of villagers' associations itself indicates a re-structuring of social relations among the varied caste groups, this being regarded as one of the important elements for modernizing the rural community.

---

14. When farmers have two or more fields, they tend to pledge the inferior one under *otti*. A similar practice is also observed in the sales tranactions.

15. This opportunity of employment for agriculture has certainly contributed at least for lessening the social unrest which would otherwise have been more serious caused by the more prevailing unemployment and underemployment in the rural communities.

So far in this section we have discussed some important factors in the recent "Green Revolution" and its socio-economic implications. The "Green Revolution" has induced rapid changes in the agricultural systems and accelerated the overall change of socio-economic activities in the village communities to a great extent. Of course, these changes have been associated with more general changing conditions for the socio-economies at regional, state, and/or national levels.

The rapid rise and fall of the socio-economic situation of individual households have been the outcome of their varied responses (and adjustments) to the changing socio-economic milieu of the village communities, regardless of the exising amount of their property holdings (such as land). Thus, it is highly doubtful poor poorer"--one of well consequences of the recent "Green Revolution." Of course, one could find such cases at the individual level, but they have much to do with influences from events in the non-agricultural sector rather than the agricultural sector. When we examined the recent land transactions, there was no clear trend toward land concentration in the relatively large land-holding class (households owning 5 acres or more).[16]

In any given time, we could observe the varied degree of inequality in the distribution of landholdings among households in the village communities. More specifically, although a few of the large households (families) could be usually found in any given time, they were not necessarily the same households (or households' descendants) after after some decades.[17] This concept canbe also applied to all categories defined by the size of landholdings.

As already pointed out in the early part of this section of the chapter, it is possible that three so-called "Green Revolutions" may have occurred during the period 1864-1980. The rapid changes in the socio-economies of the village communities during the latest "Green Revolution" were assumed to have occurred also in the other two "Green Revolutions." Including these "revolutionary" periods, continual change appears to have occurred in any given period of years, as exemplified by the changes in the *Reddiars*' percentage holding of agricultural land or in the residential areas in the studied village.

---

16. Some households belonged to this category of landholding have tried to invest their capital in non-farming business, but only a few of them have been successful.

17. Indeed, in the 1930s, the biggest landlord family in the studied village owned about 100 acres of wet land, and in 1980, none of its descendants' households had more than 15 acres.

Further changes in agriculture and village social structures in the immediate future can be expected to take place at a more rapid pace than ever before; such changes preferably should proceed in a controlled manner, thereby breaking free of the still-remaining social exploitation and economic stagnation in which they have been locked.

BIBLIOGRAPHY

Athreya, Venkotesh B.; Boklin, Gustav; Djurfeldt, Goran; and Lindberg, Staffan. *A Study of Production Relations in Agriculture, Midterm Report*. Tiruchy: Madras Institute of Development Studies and Lund University, December 1979. Mimeographed.

Geertz, Clifford. *Agricultural Involution: the Process of Ecological Change in Indonesia*. Berkeley: University of California Press, 1966.

Grist, D. H. *Rice*. London: Longmans Green, 1959.

Government of India. *Descriptive Memoir of Peruvalanallur Village of the Lalgudi Taluk of Trichinopoly District* [the First Re-settlement Register]. 1864.

Government of India. *Descriptive Memoir of Peruvalanallur Village of the Lalgudi Taluk of the Trichinopoly District* [the First Re-settlement Register]. 1898.

Government of India. *Descriptive Memoir of Peruvalanallur Village of the Lalgudi Taluk of the Trichinopoly District* [the Second Re-settlement Register]. 1927.

Government of Tamil Nadu. *Rainfall Statistics of Tamil Nadu for 30 years*. [from 1935-36 to 1964-65.] 1976.

Government of Tamil Nadu. *Rainfall Statistics of Tamil Nadu for 10* Years [from 1965-66 to 1974-75]. 1978.

Government of Tamil Nadu. *Statistical Hand Book Tamilnadu*. 1980.

Government of Madras. *The Madras Cultivating Tenants Protection Act, 1955 (As Modified up to 30th September 1965)*. 1966.

Government of Tamil Nadu. *The Madras Cultivating Tenants Acts, 1956 (As Modified up to the 15th August, 1969)*. 1971.

Government of Tamil Nadu. *The Tamil Nadu Agricultural Lands Record of Tenancy Rights Act, 1969 (As Modified up to 15th March 1977)*. 1977.

Government of Tamil Nadu. *The Tamil Nadu Irrigation Act, 1955 (As Modified up to the 31st of October 1980)*. 1981.

Gough, Kathleen. *Rural Society in Southeast India*. Cambridge: Cambridge University Press, 1981.

Hara, Tadahiko. "Esanakorai: A Study of a Village in Wet Land." *Studies in Socio-cultural Change in Rural Villages in Tiruchirappalli District, Tamil Nadu, India,* no. 2. Tokyo: Institute for Study of Languages and Cultures of Asia and Africa [ISLCAA], 1981, pp. 21-96.

Hara, Tadahiko. "Tamiru noson no shakai henyo: tiruchirappalli ken no jirei (A study of Socio-economic Change in the Villages in Lalgudi Taluk, Tiruchirappalli District, Tamil Nadu)." *Japanese Journal of Ethnology* 46 (1981): 137-172.

Humi Staff Reporter, "Changing Patterns in Land Ownership." *The Hindu* (a newspaper), January 19, 1982.

Karashima, Noboru. "Land Revenue Assessment in Cola Times as seen in the Inscriptions of the Thanjavur and Gangaikondascholapuram Temples." *studies in Socio-cultral Change in Rural Villages in Tiruchirapalli District, Tamilnadu, India,* no. 1. Tokyo: ISLCAA, 1980, pp. 35-50.

Karashima, Noboru, and Subbarayalu, Y. "Varangai/Idangai, Kaniyalar, and Irajagarattar: Social Conflict in Tamilnadu in the 15th Century." *Socio-cultural Change in Villages in Tiruchirapalli India, Part I: Pre-modern Period* (1983). pp. 133-160.

Komoguchi, Yoshimi. "Rural Community and Agriculture in the Cauvery River Basin: A Case Study of the Peruvalanallur Village of the Tiruchirappalli District, Tamil Nadu, India, Part One." *Studies in Socio-cultural Change in Rural Villages in Tiruchirappalli District, Tamil Nadu, India,* no. 2. Tokyo: ISLCAA, 1981, pp. 85-136.

Komoguchi, Yoshimi, "Rural Community and Agriculture in Bangladesh: An Essay on Three Selected Villages, Part One." *Science Reports of Geography,* no. 18. Tokyo: Komagawa University, Department of Geography, 1982, pp. 61-102.

Komoguchi, Yoshimi, "Rural Community and Agriculture in the Cauvery River Basin: A Case Study of the Peruvalanallur Village of the Tiruchirappalli District, Tamil Nadu, India, Part Two." *Science Reports of Geography,* no. 20, Tokyo: Komazawa University, 1984, pp. 1-59.

Mencher, Joan P. "Conflicts and Contradictions in the 'Green Revolution': The Case of Tamil Nadu." *Economic and Political Weekly,* February 1974, pp. 309-323.

Mencher, Joan P. *Agricultural and Social Structure in Tamil Nadu, Past Origins, Present Transformations and Future Prospects.* New Delhi: Allied Publishers Private Private Ltd, 1978.

Mizushima, Tsukasa, and Nara, Tsuyoshi. "Social Change in a Dry Village in South India." *Studies in Socio-cultural Change in Rural Villages in Tiruchirapalli District in Tamilnadu, India,* no. 4. Tokyo: ISLCAA, 1981, pp. 97-164.

Mizushima, Tsukasa. "Changes, Chances, and Choices: The Perspectives of Indian Villagers." *Socio-cultural Change in Villages in Tiruchirapalli District, Tamilnadu, India,* Part 2, Modern Period, no. 1. Tokyo: ISLCAA, 1984, pp. 37-221.

Omvedt, Gail. "Capitalist Agriculture and Rural Classes in India." *Economic and Political Weekly,* December 1981, pp. 141-159.

Patnaik, Utsa, "Class Differentiation within the Peasantry: An Approach to Analysis of Indian Agriculture." *Economic and Political Weekly,* September 1976, pp. 82-101.

Rao, G. Venkoba. *Tamil Nadu Land Reforms.* Madras: Madras Law Journal Press, 1975.

Sonachalam, K.S. *Land Reforms in Tamil Nadu.* New Delhi: Oxford and IBH Pub. Co., 1970.

Subbiah, Shanmugan P. "Rural Base in a South Indian Village: A Study into its Structural and Spatial Patterns in Mahizambadi Village of Tamil Nadu." *Studies in Socio-cultural Change in Rural Villages in Tiruchirapalli District, Tamil Nadu, India,* no. 4. Tokyo: ISLCAA, 1981, pp. 1-96.

Venkataramani, G. *Land Reform in Tamil Nadu.* Madras: Madras Institute of Development Studies, 1973.

Yamey, B.S. "The Study of Peasant Economic System: Some Concluding Comments and Questions." In Raymond Firth and B.S. Yamey, *Capital Saving and Credit in Peasant Society.* London: George Allen & Unwin Ltd., 1963. Pp. 376-386.

THE UNIVERSITY OF CHICAGO
DEPARTMENT OF GEOGRAPHY
RESEARCH PAPERS (Lithographed, 6×9 inches)

LIST OF TITLES IN PRINT

48. BOXER, BARUCH. *Israeli Shipping and Foreign Trade.* 1957. 162 p.
56. MURPHY, FRANCIS C. *Regulating Flood-Plain Development.* 1958. 216 pp.
62. GINSBURG, NORTON, editor. *Essays on Geography and Economic Development.* 1960. 173 p.
71. GILBERT, EDMUND WILLIAM. *The University Town in England and West Germany.* 1961. 79 p.
72. BOXER, BARUCH. *Ocean Shipping in the Evolution of Hong Kong.* 1961. 108 p.
91. HILL, A. DAVID. *The Changing Landscape of a Mexican Municipio, Villa Las Rosas, Chiapas.* 1964. 121 p.
97. BOWDEN, LEONARD W. *Diffusion of the Decision To Irrigate: Simulation of the Spread of a New Resource Management Practice in the Colorado Northern High Plans.* 1965. 146 pp.
98. KATES, ROBERT W. *Industrial Flood Losses: Damage Estimation in the Lehigh Valley.* 1965. 76 pp.
101. RAY, D. MICHAEL. *Market Potential and Economic Shadow: A Quantitative Analysis of Industrial Location in Southern Ontario.* 1965. 164 p.
102. AHMAD, QAZI. *Indian Cities: Characteristics and Correlates.* 1965. 184 p.
103. BARNUM, H. GARDINER. *Market Centers and Hinterlands in Baden-Württemberg.* 1966. 172 p.
105. SEWELL, W. R. DERRICK, *et al. Human Dimensions of Weather Modification.* 1966. 423 p.
107. SOLZMAN, DAVID M. *Waterway Industrial Sites: A Chicago Case Study.* 1967. 138 p.
108. KASPERSON, ROGER E. *The Dodecanese: Diversity and Unity in Island Politics.* 1967. 184 p.
109. LOWENTHAL, DAVID, editor, *Environmental Perception and Behavior.* 1967. 88 p.
112. BOURNE, LARRY S. *Private Redevelopment of the Central City, Spatial Processes of Structural Change in the City of Toronto.* 1967. 199 p.
113. BRUSH, JOHN E., and GAUTHIER, HOWARD L., JR., *Service Centers and Consumer Trips: Studies on the Philadelphia Metropolitan Fringe.* 1968. 182 p.
114. CLARKSON, JAMES D., *The Cultural Ecology of a Chinese Village: Cameron Highlands, Malaysia.* 1968. 174 p.
115. BURTON, IAN, KATES, ROBERT W., and SNEAD, RODMAN E. *The Human Ecology of Coastal Flood Hazard in Megalopolis.* 1968, 196 p.
117. WONG, SHUE TUCK, *Perception of Choice and Factors Affecting Industrial Water Supply Decisions in Northeastern Illinois.* 1968. 93 p.
118. JOHNSON, DOUGLAS L. *The Nature of Nomadism: A Comparative Study of Pastoral Migrations in Southwestern Asia and Northern Africa.* 1969. 200 p.
119. DIENES, LESLIE. Locational Factors and Locational Developments in the Soviet Chemical Industry. 1969. 262 p.
120. MIHELIČ, DUŠAN. *The Political Element in the Port Geography of Trieste.* 1969. 104 p.
121. BAUMANN, DUANE D. *The Recreational Use of Domestic Water Supply Reservoirs: Perception and Choice.* 1969. 125 p.
122. LIND, AULIS O. *Coastal Landforms of Cat Island, Bahamas: A Study of Holocene Accretionary Topography and Sea-Level Change.* 1969. 156 p.
123. WHITNEY, JOSEPH B. R. *China: Area, Administration and Nation Building.* 1970. 198 p.
124. EARICKSON, ROBERT. *The Spatial Behavior of Hospital Patients: A Behavioral Approach to Spatial Interaction in Metropolitan Chicago.* 1970. 138 p.
125. DAY, JOHN CHADWICK. *Managing the Lower Rio Grande: An Experience in International River Development.* 1970. 274 p.
126. MACIVER, IAN. *Urban Water Supply Alternatives: Perception and Choice in the Grand Basin Ontario.* 1970. 178 p.
127. GOHEEN, PETER G. *Victorian Toronto, 1850 to 1900: Pattern and Process of Growth.* 1970. 278 p.
128. GOOD, CHARLES M. *Rural Markets and Trade in East Africa.* 1970. 252 p.
129. MEYER, DAVID R. *Spatial Variation of Black Urban Households.* 1970. 127 p.
130. GLADFELTER, BRUCE G. *Meseta and Campiña Landforms in Central Spain: A Geomorphology of the Alto Henares Basin.* 1971. 204 p.
131. NEILS, ELAINE M. *Reservation to City: Indian Migration and Federal Relocation.* 1971. 198 p.
132. MOLINE, NORMAN T. *Mobility and the Small Town, 1900–1930.* 1971. 169 p.

133. SCHWIND, PAUL J. *Migration and Regional Development in the United States.* 1971. 170 p.

134. PYLE, GERALD F. *Heart Disease, Cancer and Stroke in Chicago: A Geographical Analysis with Facilities, Plans for 1980.* 1971. 292 p.

135. JOHNSON, JAMES F. *Renovated Waste Water: An Alternative Source of Municipal Water Supply in the United States.* 1971. 155 p.

136. BUTZER, KARL W. *Recent History of an Ethiopian Delta: The Omo River and the Level of Lake Rudolf.* 1971. 184 p.

139. MCMANIS, DOUGLAS R. *European Impressions of the New England Coast, 1497–1620.* 1972. 147 p.

140. COHEN, YEHOSHUA S. *Diffusion of an Innovation in an Urban System: The Spread of Planned Regional Shopping Centers in the United States, 1949–1968,* 1972. 136 p.

141. MITCHELL, NORA. *The Indian Hill-Station: Kodaikanal.* 1972. 199 p.

142. PLATT, RUTHERFORD H. *The Open Space Decision Process: Spatial Allocation of Costs and Benefits.* 1972. 189 p.

143. GOLANT, STEPHEN M. *The Residential Location and Spatial Behavior of the Elderly: A Canadian Example.* 1972. 226 p.

144. PANNELL, CLIFTON W. *T'ai-chung, T'ai-wan: Structure and Function.* 1973. 200 p.

145. LANKFORD, PHILIP M. *Regional Incomes in the United States, 1929–1967: Level, Distribution, Stability, and Growth.* 1972. 137 p.

146. FREEMAN, DONALD B. *International Trade, Migration, and Capital Flows: A Quantitative Analysis of Spatial Economic Interaction.* 1973. 201 p.

147. MYERS, SARAH K. *Language Shift Among Migrants to Lima, Peru.* 1973. 203 p.

148. JOHNSON, DOUGLAS L. *Jabal al-Akhdar, Cyrenaica: An Historical Geography of Settlement and Livelihood.* 1973. 240 p.

149. YEUNG, YUE-MAN. *National Development Policy and Urban Transformation in Singapore: A Study of Public Housing and the Marketing System.* 1973. 204 p.

150. HALL, FRED L. *Location Criteria for High Schools: Student Transportation and Racial Integration.* 1973. 156 p.

151. ROSENBERG, TERRY J. *Residence, Employment, and Mobility of Puerto Ricans in New York City.* 1974. 230 p.

152. MIKESELL, MARVIN W., editor. *Geographers Abroad: Essays on the Problems and Prospects of Research in Foreign Areas.* 1973. 296 p.

153. OSBORN, JAMES F. *Area, Development Policy, and the Middle City in Malaysia.* 1974. 291 p.

154. WACHT, WALTER F. *The Domestic Air Transportation Network of the United States.* 1974. 98 p.

155. BERRY, BRIAN J. L., *et al. Land Use, Urban Form and Environmental Quality.* 1974. 440 p.

156. MITCHELL, JAMES K. *Community Response to Coastal Erosion: Individual and Collective Adjustments to Hazard on the Atlantic Shore.* 1974. 209 p.

157. COOK, GILLIAN P. *Spatial Dynamics of Business Growth in the Witwatersrand.* 1975. 144 p.

159. PYLE, GERALD F. *et al. The Spatial Dynamics of Crime.* 1974. 221 p.

160. MEYER, JUDITH W. *Diffusion of an American Montessori Education.* 1975. 97 p.

161. SCHMID, JAMES A. *Urban Vegetation: A Review and Chicago Case Study.* 1975. 266 p.

162. LAMB, RICHARD F. *Metropolitan Impacts on Rural America.* 1975. 196 p.

163. FEDOR, THOMAS STANLEY. *Patterns of Urban Growth in the Russian Empire during the Nineteenth Century.* 1975. 245 p.

164. HARRIS, CHAUNCY D. *Guide to Geographical Bibliographies and Reference Works in Russian or on the Soviet Union.* 1975. 478 p.

165. JONES, DONALD W. *Migration and Urban Unemployment in Dualistic Economic Development.* 1975. 174 p.

166. BEDNARZ, ROBERT S. *The Effect of Air Pollution on Property Value in Chicago.* 1975. 111 p.

167. HANNEMANN, MANFRED. *The Diffusion of the Reformation in Southwestern Germany, 1518–1534.* 1975. 248 p.

168. SUBLETT, MICHAEL D. *Farmers on the Road. Interfarm Migration and the Farming of Noncontiguous Lands in Three Midwestern Townships. 1939–1969.* 1975. 228 pp.

169. STETZER, DONALD FOSTER. *Special Districts in Cook County: Toward a Geography of Local Government.* 1975. 189 pp.

170. EARLE, CARVILLE V. *The Evolution of a Tidewater Settlement System: All Hallow's Parish, Maryland, 1650–1783.* 1975. 249 pp.

171. SPODEK, HOWARD. *Urban-Rural Integration in Regional Development: A Case Study of Saurashtra, India—1800–1960.* 1976. 156 pp.

172. COHEN, YEHOSHUA S. and BERRY, BRIAN J. L. *Spatial Components of Manufacturing Change.* 1975. 272 pp.

173. HAYES, CHARLES R. *The Dispersed City: The Case of Piedmont, North Carolina.* 1976. 169 pp.

174. CARGO, DOUGLAS B. *Solid Wastes: Factors Influencing Generation Rates.* 1977. 112 pp.

175. GILLARD, QUENTIN. *Incomes and Accessibility. Metropolitan Labor Force Participation, Commuting, and Income Differentials in the United States, 1960–1970.* 1977. 140 pp.

176. MORGAN, DAVID J. *Patterns of Population Distribution: A Residential Preference Model and Its Dynamic.* 1978. 216 pp.

177. STOKES, HOUSTON H.; JONES, DONALD W. and NEUBURGER, HUGH M. *Unemployment and Adjustment in the Labor Market: A Comparison between the Regional and National Responses.* 1975. 135 pp.

179. HARRIS, CHAUNCY D. *Bibliography of Geography. Part I. Introduction to General Aids.* 1976. 288 pp.

180. CARR, CLAUDIA J. *Pastoralism in Crisis. The Dasanetch and their Ethiopian Lands.* 1977. 339 pp.

181. GOODWIN, GARY C. *Cherokees in Transition: A Study of Changing Culture and Environment Prior to 1775.* 1977. 221 pp.

182. KNIGHT, DAVID B. *A Capital for Canada: Conflict and Compromise in the Nineteenth Century.* 1977. 359 pp.

183. HAIGH, MARTIN J. *The Evolution of Slopes on Artificial Landforms: Blaenavon, Gwent.* 1978. 311 pp.

184. FINK, L. DEE. *Listening to the Learner. An Exploratory Study of Personal Meaning in College Geography Courses.* 1977. 200 pp.

185. HELGREN, DAVID M. *Rivers of Diamonds: An Alluvial History of the Lower Vaal Basin.* 1979. 399 pp.

186. BUTZER, KARL W., editor. *Dimensions of Human Geography: Essays on Some Familiar and Neglected Themes.* 1978. 201 pp.

187. MITSUHASHI, SETSUKO. *Japanese Commodity Flows.* 1978. 185 pp.

188. CARIS, SUSAN L. *Community Attitudes toward Pollution.* 1978. 226 pp.

189. REES, PHILIP M. *Residential Patterns in American Cities, 1960.* 1979. 424 pp.

190. KANNE, EDWARD A. *Fresh Food for Nicosia.* 1979. 116 pp.

192. KIRCHNER, JOHN A. *Sugar and Seasonal Labor Migration: The Case of Tucumán, Argentina.* 1980. 158 pp.

193. HARRIS, CHAUNCY D. and FELLMANN, JEROME D. *International List of Geographical Serials, Third Edition, 1980.* 1980. 457 p.

194. HARRIS, CHAUNCY D. *Annotated World List of Selected Current Geographical Serials, Fourth, Edition. 1980.* 1980. 165 p.

195. LEUNG, CHI-KEUNG. *China: Railway Patterns and National Goals.* 1980. 235 p.

196. LEUNG, CHI-KEUNG and GINSBURG, NORTON S., eds. *China: Urbanization and National Development.* 1980. 280 p.

197. DAICHES, SOL. *People in Distress: A Geographical Perspective on Psychological Well-being.* 1981, 199 p.

198. JOHNSON, JOSEPH T. *Location and Trade Theory: Industrial Location, Comparative Advantage, and the Geographic Pattern of Production in the United States.* 1981. 107 p.

199-200. STEVENSON, ARTHUR J. *The New York-Newark Air Freight System.* 1982. 440 p.

201. LICATE, JACK A. *Creation of a Mexican Landscape: Territorial Organization and Settlement in the Eastern Puebla Basin, 1520–1605.* 1981. 143 p.

202. RUDZITIS, GUNDARS. *Residential Location Determinants of the Older Population.* 1982. 117 p.

203. LIANG, ERNEST P. *China: Railways and Agricultural Development, 1875–1935.* 1982. 186 p.

204. DAHMANN, DONALD C. *Locals and Cosmopolitans: Patterns of Spatial Mobility during the Transition from Youth to Early Adulthood.* 1982. 146 p.

205. FOOTE, KENNETH E. *Color in Public Spaces: Toward a Communication-Based Theory of the Urban Built Environment.* 1983. 153 p.

206. HARRIS, CHAUNCY D. *Bibliography of Geography. Part II: Regional. Vol. 1. The United States of America.* 1984. 178 p.

207-208. WHEATLEY, PAUL. *Nāgara and Commandery: Origins of the Southeast Asian Urban Traditions.* 1983. 473 p.

209. SAARINEN, THOMAS F.; SEAMON, DAVID; and SELL, JAMES L., eds. *Environmental Perception and Behavior: An Inventory and Prospect.* 1984. 263 p.

210. WESCOAT, JAMES L., JR. *Integrated Water Development: Water Use and Conservation Practice in Western Colorado.* 1984. 239 p.

211. DEMKO, GEORGE J., and FUCHS, ROLAND J., eds. *Geographical Studies on the Soviet Union: Essays in Honor of Chauncy D. Harris.* 1984. 294 p.

212. HOLMES, ROLAND C. *Irrigation in Southern Peru: The Chili Basin.* 1986. 191 p.

213. EDMONDS, RICHARD L. *Northern Frontiers of Qing China and Tokugawa Japan: A Comparative Study of Frontier Policy.* 1985. 155 p.

214. FREEMAN, DONALD B., and NORCLIFFE, GLEN B. *Rural Enterprise in Kenya: Development and Spatial Organization of the Nonfarm Sector.* 1985. 180 p.

215. COHEN, YEHOSHUA S., and SHINAR, AMNON. *Neighborhoods and Friendship Networks: A Study of Three Residential Neighborhoods in Jerusalem.* 1985. 129 p.

217-218. CONZEN, MICHAEL P., ED. *World Patterns of Modern Urban Change: Essays in Honor of Chauncy D. Harris.* 1986.

219. KOMOGUCHI, YOSHIMI. *Agricultural Systems in the Tamil Nadu: A Case Study of Peruvalanallur Village.* 1986. 171 p.

220. GINSBURG, NORTON; OBORN, JAMES; and BLANK, GRANT. *Geographic Perspectives on the Wealth of Nations.* 1986. 131 p.